T0324730

Beyond the Triangle:
Brownian Motion, Ito Calculus, and Fokker–Planck Equation — Fractional Generalizations

Beyond the Triangle:
Brownian Motion, Ito Calculus,
and Fokker–Planck Equation
— Fractional Generalizations

Sabir Umarov
University of New Haven, USA

Marjorie Hahn
Tufts University, USA

Kei Kobayashi
Fordham University, USA

World Scientific

NEW JERSEY · LONDON · SINGAPORE · BEIJING · SHANGHAI · HONG KONG · TAIPEI · CHENNAI · TOKYO

Published by

World Scientific Publishing Co. Pte. Ltd.

5 Toh Tuck Link, Singapore 596224

USA office: 27 Warren Street, Suite 401-402, Hackensack, NJ 07601

UK office: 57 Shelton Street, Covent Garden, London WC2H 9HE

Library of Congress Cataloging-in-Publication Data

Names: Umarov, Sabir, author. | Hahn, Marjorie G., author. | Kobayashi, Kei (Mathematics professor), author.

Title: Beyond the triangle : Brownian motion, Ito calculus, and Fokker-Planck equation :
    fractional generalizations / by Sabir Umarov (University of New Haven, USA),
    Marjorie Hahn (Tufts University, USA), Kei Kobayashi (Fordham University, USA).

Description: New Jersey : World Scientific, 2017. | Includes bibliographical references and index.

Identifiers: LCCN 2017042756 | ISBN 9789813230910 (hardcover : alk. paper)

Subjects: LCSH: Brownian motion processes. | Fokker-Planck equation. | Stochastic differential equations.

Classification: LCC QA274.75 .U43 2017 | DDC 515/.353--dc23

LC record available at https://lccn.loc.gov/2017042756

British Library Cataloguing-in-Publication Data

A catalogue record for this book is available from the British Library.

For any available supplementary material, please visit
http://www.worldscientific.com/worldscibooks/10.1142/10734#t=suppl

Printed in Singapore

To our teachers

Preface

This book is devoted to the fundamental relationship between three objects: a stochastic process, stochastic differential equations driven by that process, and their associated Fokker–Planck–Kolmogorov (FPK) equations. The book contains recent results obtained in this direction. In the simplest case this is the relationship between Brownian motion with a drift, the associated stochastic differential equation (SDE) in the sense of Itô, and the classic Fokker–Planck equation. The notion of Brownian motion was coined by Albert Einstein in one of his 1905 *Annus Mirabilis* papers[1]. In this paper he gave a theoretical explanation of Brownian motion. A little earlier (in 1900) Louis Bachelier published his doctoral dissertation[2] modeling Brownian motion from the economics point of view. This work serves as the origin of the modern financial mathematics. Independently of Einstein, Marian von Smoluchowski, in his 1906 studies on fluctuating particles, provided an explanation of diffusion processes in the kinetic theory of gases. In 1908 Langevin published his work with a stochastic differential equation which was "understood mathematically" only after a stochastic calculus was introduced by Itô in 1944–48. The Fokker–Planck equation, a deterministic form of describing the dynamics of a random process in terms of transition probabilities, was introduced in 1913–17. Its complete "mathematical understanding" became available after the appearance of the distribution (generalized function) theory (Sobolev (1938), Schwartz (1951)) and was embodied in Kolmogorov's backward and forward equations.

Today the triple relationship between Brownian motion, Itô stochastic differential equations driven by Brownian motion and their associated Fokker-Plank–Kolmogorov partial differential equations is well known. Stochastic differential equations driven by Brownian motion significantly broaden the important role played by Brownian motion in both theory and applications. Moreover, solutions of such stochastic differential equations derive properties from having Brownian motion as the driving process. These facts motivated us and other researchers to investigate

---

[1] "Über die von der molekularkinetischen Theorie der Wärme geforderte Bewegung von in ruhenden Flüssigkeiten suspendierten Teilchen" ("On the Motion of Small Particles Suspended in a Stationary Liquid, as Required by the Molecular Kinetic Theory of Heat").
[2] "Théorie de la spéculation" ("The Theory of Speculation").

stochastic processes driven by processes other than Brownian motion. In this book we develop a paradigm that embodies three objects fundamental to understanding these properties and which we think makes the properties more easily accessible to scientists, economists and other researchers.

This book discusses wide fractional generalizations of this fundamental triple relationship where the driving process represents a time-changed stochastic process, the Fokker–Planck–Kolmogorov equation involves time-fractional order derivatives and spatial pseudo-differential operators, and the associated stochastic differential equation describes the stochastic behavior of the solution process. In a paradigm to be developed in this book the driving process is a key point. Driving processes to be considered include semimartingales (e.g. Brownian motion, Lévy processes), non-semimartingales (e.g. fractional Brownian motion, general Gaussian processes), and their time-changed versions. Depending on the driving process, fractional order Fokker–Planck–Kolmogorov equations and their associated stochastic differential equations describe complex stochastic processes. The complexity includes phenomena such as the presence of weak or strong correlations, different sub- or super-diffusive modes and jump effects, and combination of some or all of them. The paradigm will broaden understanding of many complex real world problems and yield mathematical models that can be used in the study of such problems. The topics discussed in the present monograph can be of interest for graduate students, beginning and established theorists, as well as researchers in applied sciences.

The monograph contains seven chapters. Chapter 1 is an introduction to the paradigm of the triple relationship discussed above in the context of the fractional Fokker–Planck–Kolmogorov equation and related topics. This chapter discusses in detail the question of why we need fractional generalizations of Fokker–Planck–Kolmogorov equation and their various applications. More detailed information about the contents of chapters of the book is also provided in this introductory chapter. Chapter 2 discusses the original triple relationship between Brownian motion as the driving process, the Itô stochastic differential equations driven by Brownian motion, and the associated Fokker–Planck–Kolmogorov equations. Chapter 3 introduces the fractional order derivatives and operators that are used in our further analysis. Chapter 4 discusses pseudo-differential operators arising in the context of fractional Fokker–Planck–Kolmogorov equations. Chapters 5 and 6 develop a stochastic calculus for time-changed semimartingales and also discuss the processes' continuous time random walk (CTRW) approximants and numerical simulation schemes. Finally, in Chapter 7 we develop a theory of fractional order Fokker–Planck–Kolmogorov equations and their associated stochastic processes fully describing the paradigm given in Chapter 1. Some sections in Chapter 7 are divided into a theory subsection followed by an applications subsection. This book was motivated by the authors' published results in [Hahn et al. (2012), Hahn et al. (2011b), Hahn and Umarov (2011), Hahn et al. (2011a), Kobayashi (2011), Jum and Kobayashi (2016), Umarov (2015b)].

Acknowledgments

We are grateful to our colleagues with whom we discussed various topics included in the book in personal meetings, workshops, colloquiums, and conferences. Our deep gratitude goes to professors Rudolf Gorenflo, Virginia Kiryakova, Yury Luchko, Francesco Mainardi, Mark Meerschaert, Erkan Nane, Jan Rosiński, Jelena Ryvkina, Meredith Burr, Jamie Wolf, and Ernest Jum. We also acknowledge with gratitude the departments of mathematics at Tufts University, the University of New Haven, the University of Tennessee, and Fordham University for providing working environments conducive to completing this research.

Sabir Umarov, University of New Haven
Marjorie Hahn, Tufts University
Kei Kobayashi, Fordham University

June 30, 2017

Contents

Chapter 1

Introduction

1.1 Why fractional generalizations of the Fokker–Planck equation?

For the reader's convenience we start the book with a discussion of the importance of the fractional generalizations of the classic Fokker–Planck equation introduced in 1914 first by Adrian Fokker [Fokker (1914)] and then extended in 1917 by Max Planck [Planck (1917)] to the general case. A number of inspiring investigations preceded these two seminal works, including Bachelier's dissertation [Bachelier (1900)] on speculations over shares and Einstein's paper [Einstein (1905)] on fluctuations of a small particle suspended in a stationary liquid. In the latter Brownian motion was explained from the statistical point of view and the Fokker–Planck equation corresponding to Brownian motion was derived. One byproduct of this work was an elegant technique to calculate the number of atoms/molecules in one mole weight of substance, known as Avogadro's[1] number. Independently of Einstein, in 1906 Smoluchowski [Smoluchowski (1906)] derived the Smoluchowski equation of diffusion of gas molecules, which is indeed a particular case of the Fokker–Planck equation. Langevin's 1908 paper [Langevin (1908)] contains the first stochastic differential equation with random fluctuations. There is no doubt that the Fokker–Planck equation became a new cornerstone in statistical physics. It both expanded the scope of the domain of applications of statistical physics and enhanced its tools after the foundational works done by Boltzmann, Gibbs, Maxwell, and others.

It is not surprising that a huge number of applications of the Fokker–Planck equation in almost all branches of the modern sciences and engineering were discovered in the 20th century. Many books are written about the Fokker–Planck equation reflecting both theoretical developments and their applications; see, for example, [Risken and Frank (1996), Gardiner (1985), Frank (2006), Umarov (2015b), Shizgal (2015), McCauley (2013)]. One of the theoretical advances is a deep mathematical result obtained by Kolmogorov in 1931 on the dual nature of forward and backward Fokker–Planck (parabolic partial differential) equations associated with

---

[1]Amedeo Avogadro (1776-1856) was an Italian physicist. In 1811 he theoretically proposed that the number of molecules of any gas in the same volume is constant if the pressure and temperature are fixed.

Markovian stochastic processes [Kolmogorov (1931)][2]. Giving credit to all three authors justifies our use of the term *Fokker–Planck–Kolmogorov (FPK for short) equations* for the forward and backward Fokker–Planck equations.

Interest in fractional order FPK equations has appeared relatively recently, though fractional order differential equations were used in mathematical modeling long ago. Perhaps the first application of fractional integro-differential operators was Abel's integral equation of the first kind connected with the famous tautochrone problem published in 1826. In paper [Abel (1826)], Abel obtained a solution of this problem essentially in the form of the fractional derivative in the sense of Riemann–Liouville. Oldham and Spanier, in their 1974 book [Oldham and Spanier (1974)], discussed fractional generalizations of transport and diffusion equations, which can be interpreted as a particular case of fractional FPK equations. Starting in the 1980s, applications of fractional calculus to various fields significantly increased; see, e.g. [Nigmatullin (1986), Wyss (1986), Schneider and Wyss (1989), Fujita (1990), Schneider (1990)]. These works, as well as investigations performed later, revealed many important intrinsic properties of processes modeled by fractional equations, including hereditary properties and memory effects, oscillation-relaxation properties, connections with Lévy processes and subordinating processes, and many other properties, which cannot be captured by integer order models. More currently, intensive research in theoretical developments of fractional FPK equations and their applications in modeling various complex processes have involved novel ideas such as introducing distributed and variable fractional order differential operators [Lorenzo and Hartley (2002), Umarov and Gorenflo (2005a), Meerschaert and Scheffler (2006), Kochubei (2008), Umarov (2015b)]. It is impossible to review all the works on fractional order differential equations appearing daily at an increasing rate. However, for the theoretical background of fractional calculus, its historical development, and various applications, we can refer the reader to books [Samko et al. (1993), Podlubny (1998), Kilbas et al. (2006), Mainardi (2010), Umarov (2015b)].

In the last few decades, fractional order FPK equations have appeared as an essential tool for the study of dynamics of various complex stochastic processes arising in anomalous diffusion in physics [Metzler and Klafter (2000), Zaslavsky (2002)], finance [Gorenflo et al. (2001)], hydrology [Benson et al. (2000)], and cell biology [Saxton and Jacobson (1997)]. Complexity includes phenomena such as the presence of weak or strong correlations, different sub- or super-diffusive modes and jump effects. Consider the following example from cell biology (see details in [Saxton and Jacobson (1997)]). Based on single particle tracking, experimental studies of the motion of proteins and other macromolecules in the cell membrane show apparent subdiffusive motion, i.e. the rate at which the particles spread out is slower than that of a diffusion, such as Brownian motion. Additionally, several subdiffusive modes simultaneously affect the motion. One experiment that describes such phenomena

---

[2]This paper is a strictly mathematical treatment of the topic. Perhaps this is the reason why it does not contain citations to papers [Einstein (1905), Fokker (1914), Planck (1917)] but has a citation to [Bachelier (1900)], which is also strictly mathematical.

is provided in [Ghosh and Webb (1994)], which recorded that approximately 50% of case measurements on the LDL receptor labeled with $diILDL$ show subdiffusive motion, with subdiffusion modes whose parameters β are between 0.2 and 0.9. Here, the smaller the parameter β, the more slowly the particles scatter, whereas the case $\beta = 1$ corresponds to the classical diffusion. Examples can be drawn from numerous other fields.

1.2 The problem formulation

The theory developed in the book is represented as a paradigm of the relationship between solutions to stochastic differential equations, their associated Fokker–Planck–Kolmogorov equations, and continuous time random walk approximations of these processes, placing the driving process of the stochastic differential equations in the center. In the classic case the paradigm places Brownian motion at the center and surrounds it by a triangle providing interconnections between

(1) solution to an Itô stochastic differential equation (SDE) driven by Brownian motion,
(2) random walk (RW)-based approximants for the SDE solution, and
(3) Fokker–Planck–Kolmogorov (FPK) equations governing the dynamics of the SDE solution. The Fokker–Planck–Kolmogorov equations[3] consist of deterministic partial differential equations governing the forward and backward time evolution of the density (transition probabilities) of the SDE solution. The two FPK equations have a dual relationship.

The paradigm originated with the trivial SDE: $dX_t = dB_t$, where Brownian motion B is the solution. In this case, the paradigm encompasses the deep and incredibly useful interconnections between B, its random walk approximants and the heat (diffusion) equation, an indicator of the importance of extending those kinds of connections to solutions of more complicated SDEs driven by Brownian motion. For example, for b and σ constants, consider the SDE

$$dX_t = bX_t\, dt + \sigma X_t\, dB_t,$$

with $X_0 = 1$, whose

(1) solution is geometric Brownian motion, i.e. $X_t = \exp\{(b - \sigma^2/2)t + \sigma B_t\}$,
(2) RW-based approximants are

$$X_t^n = \exp\{(b - \sigma^2/2)[Z^n, Z^n]_t + \sigma Z_t^n\}$$

with RWs $Z_t^n = \sum_{j=1}^{\lfloor nt\rfloor} \xi_i/\sqrt{n}$ approximating B_t with quadratic variation $[Z^n, Z^n]_t$, and
(3) forward FPK equation is

$$\frac{\partial u(t,x)}{\partial t} = -b\frac{\partial}{\partial x}[x\, u(t,x)] + \frac{\sigma^2}{2}\frac{\partial^2}{\partial x^2}[x^2\, u(t,x)], \quad t > 0,\ x \in \mathbb{R}.$$

---

[3]Scientists tend to use the terminology "Fokker–Planck equation" for the forward FPK equation.

Our objective is much broader. We seek to develop a useful paradigm for SDE solutions when the driving process comes from classes, such as Lévy processes, fractional Brownian motions, general Gaussian processes and time-changed versions of these classes of processes (i.e. those with an embedded nondecreasing random time-change). Towards this end, the general paradigm pictured in Figure 1.1 below shows the driving process (DP) in the center surrounded by the three objects whose identifications and interconnections are sought:

[[SDE]] denotes the solution of the type of SDE appearing in that box,
[[CTRW]] designates approximants for the SDE solution based on continuous time random walk (CTRW) approximants for the driving process, and
[[FPK]] designates the type of differential equation/operator giving the time-evolution of the probability densities for the SDE solution in the FPK equation.

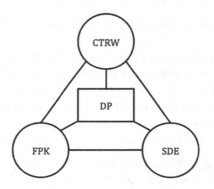

Fig. 1.1   General paradigm

There are scientific benefits to the results of such a paradigm. Indeed, scientists, engineers, and economists find Lévy processes and fractional Brownian motions or their time-changed versions adequate models for various processes in the natural (physics, hydrology, biology, etc) and social (finance, economics) sciences. A sample of such processes and applications appears, for instance, in the monographs [Barndorf-Nielsen et al. (2001), Bertoin (1996), Mandelbrot (1997), Meerschaert and Scheffler (2001), Risken and Frank (1996), Samorodnitsky and Taqqu (1994), Sato (1999), Uchaykin and Zolotarev (1999), Umarov (2015b)], and survey papers [Bouchaud and Georges (1990), Metzler and Klafter (2000), Zaslavsky (2002)]. Furthermore, the paradigm provides a theoretical platform which facilitates and makes more widely accessible the organization, study, and application of the driving processes as well as their corresponding SDEs for both theoretical and applied researchers. In the paradigm, the [[CTRW]]–[[FPK]] and [[CTRW]]–[[SDE]] correspondences involve methods of approximation which often suggest ways a process might arise. On the other hand, the [[FPK]]–[[SDE]] correspondence is between the stochastic process solving the SDE and its transition probabilities which solve the FPK equation. Transition probabilities are valuable for simulations. At the same time, the [[FPK]]–[[SDE]] correspondence facilitates use of known properties of the solution process which are not accessible from simulations. So researchers who initially approach the evolution of a phenomenon deterministically (via the FPK equation) should find the [[FPK]]–[[SDE]] correspondence extremely valuable.

The driving process of an SDE plays a key role in the dynamics and evolution of the solution to that SDE. Processes associated with time-fractional order FPK equations are usually driven by complicated time-changed processes. Even in the simplest case of the time-fractional diffusion equation

$$\partial^\beta u = \kappa_\beta \Delta u,$$

where κ_β is the diffusion coefficient, Δ is the Laplace operator, and ∂^β is a time-fractional derivative of order $0 < \beta < 1$, the driving process is not even a Lévy process. In this specific case, the density of the associated driving process is the solution of the time-fractional diffusion equation. This indicates that more generally, understanding properties of the driving process should elucidate properties of the density. There are actually several different time-fractional derivatives of order $0 < \beta < 1$, as we will see below.

Looking at the different cases of the paradigm studied in this book from the point of view of a scientist, the classification would be according to the type of FPK put into the box, [[FPK]], in the paradigm. For example, let FPK denote the classical Fokker–Planck–Kolmogorov equation whose driving process is Brownian motion. As was noted above, the FPK equation establishes a relationship between Itô's SDE driven by Brownian motion and its associated partial differential equation. Namely, for the SDE driven by Brownian motion

$$dX_t = F(t, X_t)dt + G(t, X_t)dB_t,$$

where the drift coefficient $F(t, x)$ and the diffusion coefficient $G(t, x)$ satisfy some continuity and growth conditions, its associated FPK equation has the form

$$\frac{\partial p(t, x)}{\partial t} = \left[-\frac{\partial F(t, x)}{\partial x} + \frac{1}{2} \frac{\partial^2 G(t, x)}{\partial x^2} \right] p(t, x), \quad t > 0, \ x \in \mathbb{R}. \tag{1.1}$$

There are several different forms of fractional generalization of the FPK equation. One of the forms of the time-fractional FPK equation used frequently by physicists (see e.g. [Metzler et al. (1999), Sokolov and Klafter (2006), Lv et al. (2012), Magdziarz et al. (2014)]) is

$$\frac{\partial u(t, x)}{\partial t} = \left[-\frac{\partial F(t, x)}{\partial x} + \frac{1}{2} \frac{\partial^2 G(t, x)}{\partial x^2} \right] D_+^{1-\beta} u(t, x), \quad t > 0, \ x \in \mathbb{R}, \tag{1.2}$$

with the initial condition $u(0, x) = \delta_0(x)$. Here $D_+^{1-\beta}$ is the Riemann–Liouville time-fractional derivative of order $1 - \beta$ defined in Chapter 3, where $0 < \beta < 1$. In the limit case $\beta \to 1$, equation (1.2) recovers the classical FPK equation represented in (1.1). We will see in Chapter 7 that there is a relationship between the solution $p(t, x)$ of equation (1.1) and the solution $u(t, x)$ of equation (1.2), as well as between their associated stochastic processes. If $F(t, x) = F(x)$ and $G(t, x) = G(x)$, i.e. the coefficients do not depend on the time variable t, then one can easily verify that (1.2) is equivalent to the following version of the time-fractional FPK equation

$$D_*^\beta u(t, x) = \left[-\frac{\partial F(t, x)}{\partial x} + \frac{1}{2} \frac{\partial^2 G(t, x)}{\partial x^2} \right] u(t, x), \quad t > 0, \ x \in \mathbb{R}, \qquad (1.3)$$

where D_*^β is the Caputo time-fractional derivative of order β defined in Chapter 3. Equation (1.2) in the case $F(t, x) = F(x)$ and $G(t, x) = $ constant was first established using a CTRW approach in [Metzler et al. (1999)], in the case $F(t, x) = F(t)$ and $G(t, x) = $ constant in [Sokolov and Klafter (2006)], and in the case $F(t, x) = F(x)f(t)$ and $G(t, x) = G(x)g(t)$ in [Lv et al. (2012)]. This indicates the usefulness of CTRW approximants.

The basic building block with the driving process being a Brownian motion is first extended to a Lévy process. Lévy processes are a larger class of processes for mathematical modeling which includes Brownian motion and processes with jumps, such as the Lévy stable processes with power law tails. Brownian motion and Brownian motion with drift are the only Gaussian Lévy processes with continuous sample paths. As we will see in Chapter 5, any Lévy process in $\mathbb{R}^n$ is defined uniquely through three parameters: a drift vector $b \in \mathbb{R}^n$, a covariance matrix $\Sigma \in \mathbb{R}^{n \times n}$, and a Lévy measure μ (see definitions in Section 5.3). Lévy processes have jumps of small and large sizes of different intensity depending on the defining parameters. In graphs the latter appears in the form of finite discontinuities[4]. One can express schematically these discontinuities as segments of vertical lines (left graph in Figure 1.2).

The FPK equation for the transition probabilities of a solution to an SDE driven by a Lévy process will involve a spatial pseudo-differential operator. For discussion and properties of pseudo-differential operators, see [Taylor (1981), Hörmander (1983), Jacob (2001, 2002, 2005)]. Pseudo-differential operators connected with Lévy processes and some of their properties are presented in Chapter 4.

On the other hand, in Chapter 7, we will see that a time-fractional derivative of order $\beta \in (0, 1)$ is associated with a time-change by the inverse of a Lévy stable subordinator of index β. A generalization of a time-fractional derivative of order β is the fractional derivative of distributed orders which is a mixture of time-fractional derivatives of different values of β with respect to some mixing measure. Papers [Hahn et al. (2012), Hahn and Umarov (2011)] study time-fractional or time-distributed FPK pseudo-differential equations associated to SDEs with time-independent coefficients whose driving processes are Lévy processes time-changed by the inverse of one or a mixture of many Lévy stable subordinators with different values of β.

---

[4]'Lévy flights' in the physics literature.

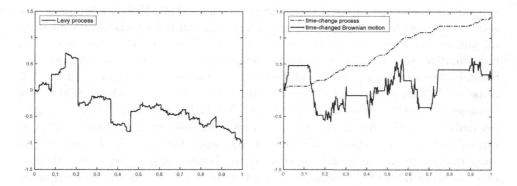

Fig. 1.2   Left Figure: Sample path of a Lévy process. Long jumps are schematically shown as vertical lines. Right Figure: Sample path of a time-changed Brownian motion in which horizontal line segments are present. The time-change process here is the inverse of a stable subordinator with stability index $\beta = 0.8$.

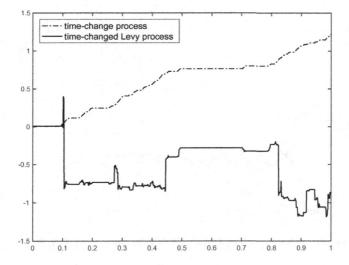

Fig. 1.3   Sample path of a time-changed Lévy process. Both vertical and horizontal line segments are present. The time-change process is the inverse of a stable subordinator with stability index $\beta = 0.8$.

Typical graphs of sample paths of time-changed stochastic processes with the inverse β-stable subordinator contain small and large segments of the horizontal line depending on the parameter β (right graph in Figure 1.2). Hence, Lévy processes subordinated with such a time-change process contain both vertical and horizontal line segments in their graphs (see Figure 1.3). Therefore, they can be used as mathematical models of various real world processes arising in financial transactions, underground water flows, cell biology, etc. For instance, Figures 1.4 and 1.5 respectively present currency exchange rates of United States Dollar (USD) versus Chinese Yuan (CNY) and Japanese Yen (JPY) during the time period January–November of 2015. We generated the figures using historical data obtained from https://www.ofx.com/en-au/forex-news/historical-exchange-rates/. In both graphs one can observe various lengths of vertical and horizontal line segments. We note that when the value of the parameter β is closer to 1, fewer horizontal line segments of large length occur. If one assumes that the stochastic processes in Figures 1.4 and 1.5 follow the model with time-changed Lévy processes, then one can expect that the value of β in Figure 1.5 is larger than its value in Figure 1.4.

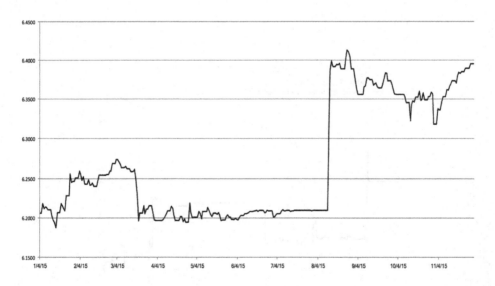

Fig. 1.4   Currency exchange rates: USD vs. CNY in the period 1/4/2015–11/30/2015.

Many of the tools available to study processes driven by Brownian motion are not available for those driven by general Gaussian or non-Gaussian Lévy processes or their time-changed versions. For instance, Lévy processes need not be martingales; they are only semimartingales [Applebaum (2009), Jacod and Shiryaev (1987)]. Fractional Brownian motions other than Brownian motion are not even semimartingales [Biagini et al. (2008), Lipster and Shiryaev (1989), Rogers (1997)] and do not possess the Markov property [Coutin and Decreusefond (1997), Mc-

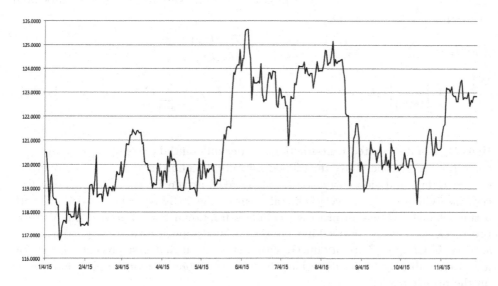

Fig. 1.5   Currency exchange rates: USD vs. JPY in the period 1/4/2015–11/30/2015.

Cauley et al. (2006)], but Lévy processes do [Applebaum (2009)]. Lévy processes and their time-changed versions are both semimartingales; hence, SDEs driven by either of them can be based on SDEs for semimartingales as will be seen in Chapter 6. However, other avenues need to be explored for defining SDEs when the driving process is a fractional Brownian motion or a Gaussian process or their time-changed versions; see Chapter 7. In order to overcome obstacles that arise, we will widely use methods and tools of stochastic integration driven by semimartingales, integral transformations, pseudo-differential and fractional order operators, continuous time random walk (CTRW) based approximants, and their various modifications.

CTRWs in the paradigm in Figure 1.1 are described by two sequences of random variables: one representing the height of the jumps; one representing the time spent waiting between successive jumps. Time-changed stochastic processes of interest in the fractional generalizations considered in this book are typically CTRW limits of triangular arrays (or scaled sequences as a special case); see Chapter 6. Time-changed Lévy processes, which often arise in applications as scaling limits of CTRWs, provide models suitable for complicated phenomena in many areas, including mathematical finance, geology, hydrology, cell biology, to name only a few. For applications, having reliable numerical approximation schemes for such time-changed processes is also important. An efficient algorithm for simulating solutions of SDEs driven by a time-changed Brownian motion is presented in Chapter 6.

Chapter 7 is a culminating chapter towards which much of the other material has been focused. As we will show in Chapter 7, for a stochastic process determined as a solution to an SDE driven by a time-changed Lévy process, in which the time-change process is the inverse to a mixture of stable subordinators, the associated FPK equation has the form

$$D_\mu v(t,x) = \sum_{j=1}^{n} b_j(x) \frac{\partial v(t,x)}{\partial x_j} + \sum_{i,j=1}^{n} \sigma_{i,j}(x) \frac{\partial^2 v(t,x)}{\partial x_i \partial x_j}$$

$$+ \int_{\mathbb{R}^n \setminus \{0\}} \left[v(t, x + G(x,w)) - v(t,x) - \boldsymbol{I}_{\{|w|<1\}} \sum_{j=1}^{n} G_j(x,w) \frac{\partial v(t,x)}{\partial x_j} \right] \nu(dw),$$

$$t > 0, \ x \in \mathbb{R}^n.$$

Here D_μ is a distributed time-fractional order differential operator (see Chapter 3) corresponding to the above mentioned mixture of subordinators, and the operator on the right hand side corresponds to the SDE driven by the original Lévy process (see Chapter 5). A very general form of the FPK equation associated with a time-changed Gaussian process (including fractional Brownian motion, Volterra processes, etc.) is also derived in Chapter 7. Since applications are important, some sections in Chapter 7 are organized with a subsection that first presents theory followed immediately by a subsection which provides one or more applications based on the prior material.

Time-fractional FPK equations form a subclass of time-fractional order pseudo-differential equations with non-smooth symbols. The existence and uniqueness of solutions of initial and boundary value problems for general fractional order pseudo-differential equations as well as the function spaces to which these solutions belong are studied in the book [Umarov (2015b)].

In many applied problems arising in such areas as plasma physics, many-body systems, astrophysics, biophysics, nonextensive statistical mechanics, nonlinear hydrodynamics, neurophysics, etc., the drift coefficient F and/or the diffusion coefficient G in equation (1.1) may depend also on u. In such cases, the FPK equation is nonlinear. In this book we do not consider nonlinear FPK equations. We refer the interested reader to the book [Frank (2006)] for (non-fractional) nonlinear FPK equations and to the book [Tsallis (2009)] for fractional order nonlinear FPK equations.

Chapter 2

The original triangle: Brownian motion, Itô stochastic calculus, and Fokker–Planck–Kolmogorov equation

2.1 Introduction

In this chapter we first summarize definitions and basic properties of Brownian motion and Itô's stochastic integrals, followed by a discussion on a derivation of the Fokker–Planck–Kolmogorov (FPK) equation. The triple relationship among Brownian motion, Itô's calculus, and FPK equation will be extended in later chapters to a general framework containing various stochastic processes, their stochastic calculus, and associated FPK equations.

2.2 Brownian motion

This section is devoted to a brief discussion on Brownian motion, one of the most fundamental objects in probability theory whose first mathematical construction was due to Norbert Wiener in 1923. A one-dimensional stochastic process $B = (B_t)_{t \geqslant 0}$ on a probability space $(\Omega, \mathcal{F}, \mathbb{P})$ is called a *Brownian motion* if (B_t) has independent and stationary increments and $B_t \sim N(0, t)$ for each $t > 0$, where $N(\mu, a)$ is a normal distribution with mean μ and variance a. This implies that for $0 = t_0 < t_1 < \cdots < t_n < \infty$, the random variables $B_{t_j} - B_{t_{j-1}}, 1 \leqslant j \leqslant n$, are independent and $B_{t_j} - B_{t_{j-1}} \sim N(0, t_j - t_{j-1})$. A continuous modification of (B_t) exists due to Kolmogorov's continuity criterion. *Throughout this book, unless mentioned otherwise, Brownian motion is assumed to have continuous sample paths and start at zero with probability one.*

A Brownian motion appears as a scaling limit of random walks. Namely, let Y_i, $i = 1, 2, 3, \ldots$ be a sequence of i.i.d. random variables with mean 0 and variance 1, and let $S_0 = 0$ and $S_n = \sum_{i=1}^{n} Y_i$ for $n \geqslant 1$. We use the notation $(S_t)_{t \geqslant 0}$ to denote the continuous-time process obtained by linearly interpolating the discrete-time process $(S_n)_{n \in \{0\} \cup \mathbb{N}}$. Then Donsker's invariance principle states that the sequence $(n^{-1/2} S_{nt})_{t \geqslant 0}$, $n = 1, 2, 3, \ldots$, converges weakly to a Brownian motion, where the convergence occurs in the space $C[0, \infty)$ of continuous functions on $[0, \infty)$ with the metric $d(\omega_1, \omega_2) = \sum_{n=1}^{\infty} \frac{1}{2^n} \max_{0 \leqslant t \leqslant n}(|\omega_1(t) - \omega_2(t)| \wedge 1)$ with $a \wedge b = \min\{a, b\}$.

One of the properties that makes Brownian motion difficult yet interesting to analyze is that its sample paths are nowhere differentiable and have infinite first variation; i.e. for a fixed $t > 0$,

$$\sup_{\Pi} \sum_{i=1}^{n} |B_{t_i} - B_{t_{i-1}}| = \infty \quad \text{a.s.},$$

where the supremum is taken over all partitions $\Pi = \{0 = t_0 < t_1 < \cdots < t_n = t\}$ of the interval $[0, t]$. One implication of having infinite first variation is that Brownian sample paths cannot be used as integrators of Lebesgue-Stieltjes integrals. Kiyoshi Itô overcame this difficulty in the 1940s and developed what is now called Itô's integration. Two components of Brownian motion are fundamental in a discussion of Itô's calculus: the quadratic variation and martingale property. Namely, the *quadratic variation* of a given stochastic process (X_t) on an interval $[0, t]$ is defined to be

$$[X, X]_t = \lim_{\|\Pi_n\| \to 0} \sum_{i=1}^{n} |X_{t_i} - X_{t_{i-1}}|^2, \tag{2.1}$$

where $\Pi_n = \{0 = t_0 < t_1 < \cdots < t_n = t\}$ is a partition of $[0, t]$ and the limit is defined via convergence in probability as $\|\Pi_n\| = \max_{1 \leqslant i \leqslant n}(t_i - t_{i-1}) \to 0$. Even though Brownian motion does not have finite first variation, it does have finite quadratic variation. In fact, for $t \geqslant 0$,

$$[B, B]_t = t,$$

where the convergence in (2.1) occurs a.s. if $\|\Pi_n\| = o(1/\log n)$ (see [Dudley (1973)]). Let $(\mathcal{F}_t)_{t \geqslant 0}$ be a filtration on $(\Omega, \mathcal{F}, \mathbb{P})$; it is an increasing family of sub-σ-algebras of $\mathcal{F}$. By definition, an integrable stochastic process $(M_t)_{t \geqslant 0}$ is an $(\mathcal{F}_t)$-*martingale* if it is $(\mathcal{F}_t)$-adapted (i.e. each X_t is $\mathcal{F}_t$-measurable) and for all $0 \leqslant s \leqslant t < \infty$,

$$\mathbb{E}[M_t | \mathcal{F}_s] = M_s \quad \text{a.s.}$$

Brownian motion is a martingale with respect to the filtration generated by itself. Namely, (B_t) is an $(\mathcal{F}_t)$-martingale, where

$$\mathcal{F}_t = \sigma(\sigma(B_s; 0 \leqslant s \leqslant t), \mathcal{N})$$

with $\mathcal{N}$ denoting all the $\mathbb{P}$-null sets. The addition of $\mathcal{N}$ guarantees that the filtration is right-continuous and contains all $\mathbb{P}$-null sets in $\mathcal{F}$.

A generalization of Brownian motion to a higher dimension is straightforward. An m-*dimensional Brownian motion* is a stochastic process $(B_t^1, \ldots, B_t^m)_{t \geqslant 0}$ with the components being independent one-dimensional Brownian motions. It is a martingale with respect to the filtration generated by the Brownian motion, and it has *quadratic covariation* (or *cross variation*)

$$[B^k, B^\ell]_t = \lim_{\|\Pi\| \to 0} \sum_{i=1}^{n} (B_{t_i}^k - B_{t_{i-1}}^k)(B_{t_i}^\ell - B_{t_{i-1}}^\ell) = \delta_{k,\ell}\, t,$$

where Π is a partition of $[0, t]$ and $\delta_{k,\ell}$ is the Kronecker delta.

2.3 Itô calculus

This section illustrates an idea of constructing stochastic integrals driven by a one-dimensional Brownian motion. Let $(\Omega, \mathcal{F}, \mathbb{P})$ be a complete probability space equipped with a filtration $(\mathcal{F}_t)$ satisfying the usual conditions; i.e. the filtration is right-continuous and contains all $\mathbb{P}$-null sets in $\mathcal{F}$. Let $B = (B_t)$ be a one-dimensional $(\mathcal{F}_t)$-Brownian motion. This implies that B is adapted to the filtration $(\mathcal{F}_t)$ and for all $0 \leqslant s \leqslant t$, $B_t - B_s$ follows a $N(0, t-s)$ distribution and is independent of $\mathcal{F}_s$. Fix $T > 0$. A stochastic process $H : [0, T] \times \Omega \to \mathbb{R}$ is called *progressively measurable* if for each $t \in [0, T]$, the mapping $[0, t] \times \Omega \ni (s, \omega) \mapsto H_s(\omega) \in \mathbb{R}$ is measurable with respect to the product σ-algebra $\mathcal{B}[0, t] \times \mathcal{F}_t$, where $\mathcal{B}[0, t]$ is the Borel σ-algebra of the set $[0, t]$. Let a class $\mathcal{L}^2[0, T]$ consist of all progressively measurable processes H such that

$$\mathbb{E}\left[\int_0^T H_s^2 \, d[B, B]_s\right] = \mathbb{E}\left[\int_0^T H_s^2 \, ds\right] < \infty.$$

Then $\mathcal{L}^2[0, T]$ is a closed subspace of the Hilbert space $L^2([0, T] \times \Omega, dt \times d\mathbb{P})$ with the metric induced by the norm $\|H\|_T = \left(\mathbb{E}[\int_0^T H_s^2 \, ds]\right)^{1/2}$.

To define a stochastic integral of $H \in \mathcal{L}^2[0, T]$, we first confine our attention to the dense subspace $\mathcal{L}_0^2[0, T]$ of $\mathcal{L}^2[0, T]$ consisting of H such that

$$H_t(\omega) = a_0(\omega) \, \boldsymbol{I}_{\{0\}}(t) + \sum_{j=1}^{N} a_j(\omega) \, \boldsymbol{I}_{(t_{j-1}, t_j]}(t)$$

where $\{t_j\}_{j=0}^N$ is a sequence of real numbers such that

$$0 = t_0 < t_1 < t_2 < \cdots < t_N \leqslant T$$

and $(a_j)_{j=0}^N$ are bounded random variables such that a_j is $\mathcal{F}_{t_{j-1}}$-measurable for $j = 1, 2, \ldots, N$. Here, $\boldsymbol{I}_A$ denotes the indicator function over a set A. The *stochastic integral* driven by (B_t) of a process $H \in \mathcal{L}_0[0, T]$ with the above representation is defined to be

$$\int_0^t H_s \, dB_s \equiv \sum_{j=1}^{N} a_j \, (B_{t \wedge t_j} - B_{t \wedge t_{j-1}}),$$

where $t \in [0, T]$. The process $\left(\int_0^t H_s \, dB_s\right)_{t \in [0, T]}$ becomes an $(\mathcal{F}_t)$-martingale, known as the *martingale transform*. Moreover, the Itô isometry holds:

$$\mathbb{E}\left[\left(\int_0^T H_s \, dB_s\right)^2\right] = \mathbb{E}\left[\int_0^T H_s^2 \, ds\right]. \tag{2.2}$$

For a given $H \in \mathcal{L}^2[0, T]$, let $\{H^{(n)}\} \subset \mathcal{L}_0^2[0, T]$ be an approximating sequence for the process H; i.e. $\|H - H^{(n)}\|_T \to 0$. Let $I_T(H^{(n)}) = \int_0^T H_s^{(n)} \, dB_s$ for $n \geqslant 1$. Since $\{I_T(H^{(n)}); n \geqslant 1\}$ forms a Cauchy sequence in $L^2(\Omega)$, it is possible to define the following limit in the L^2 sense:

$$I_T(H) \equiv \lim_{n \to \infty} I_T(H^{(n)}).$$

This is well-defined; i.e. the limit is independent of the choice of the approximating sequence for H. Now, for each fixed $t > 0$, note that $H I_{[0,t]} \in \mathcal{L}^2[0,T]$ and that $I_T(H I_{[0,t]})$ is defined as above. There exists a continuous martingale $(I_t(H))_{t \in [0,T]}$ such that $I_t(H) = I_T(H I_{[0,t]})$ a.s. The process $(I_t(H))_{t \in [0,T]}$ is called the *stochastic integral* of $H \in \mathcal{L}^2[0,T]$ and denoted

$$I_t(H) = \int_0^t H_s \, dB_s$$

for $t \in [0,T]$. Again, the process $\left(\int_0^t H_s \, dB_s \right)_{t \in [0,T]}$ is an $(\mathcal{F}_t)$-martingale and the Itô isometry (2.2) holds.

The class of integrands $\mathcal{L}^2[0,T]$ can be extended as follows. Let $\mathcal{L}^2_{loc}[0,T]$ denote the set of all progressively measurable processes H satisfying

$$\mathbb{P}\left(\int_0^T H_t^2 \, d[B,B]_t < \infty \right) = \mathbb{P}\left(\int_0^T H_t^2 \, dt < \infty \right) = 1.$$

For $H \in \mathcal{L}^2_{loc}[0,T]$, define a sequence $\{\sigma_n\}_{n=1}^\infty$ of bounded $(\mathcal{F}_t)$-stopping times by

$$\sigma_n = n \wedge \inf\left\{ t \in [0,T] \; ; \; \int_0^t H_s^2 \, d[B,B]_s = n \right\},$$

where $a \wedge b = \min\{a,b\}$. Clearly, $\sigma_n \to \infty$ as $n \to \infty$ a.s. For each $n \geq 1$, the process $H^{(n)}$ defined by $H_t^{(n)} = H_t \, I_{[0,\sigma_n]}(t)$ is an element of $\mathcal{L}^2[0,T]$, and hence, the stochastic integral $\left(\int_0^t H_s^{(n)} \, dB_s \right)_{t \in [0,T]}$ is defined. Furthermore, for $1 \leq n \leq m$ and $t \in [0, \sigma_n \wedge T]$,

$$\int_0^t H_s^{(n)} \, dB_s = \int_0^t H_s^{(m)} \, dB_s.$$

Therefore, the process $X = (X_t)$ given by

$$X_t = \int_0^t H_s^{(n)} \, dB_s \quad \text{for} \quad t \in [0, \sigma_n \wedge T]$$

is well-defined. The process X is called the *stochastic integral* of $H \in \mathcal{L}^2_{loc}[0,T]$ and denoted $X_t = \int_0^t H_s \, dB_s$ for $t \geq 0$. It is known that X is an $(\mathcal{F}_t)$-*local martingale*; i.e. there exists a nondecreasing sequence $\{\tau_k\}$ of $(\mathcal{F}_t)$-stopping times going to infinity such that $(X_{\tau_k \wedge t})_{t \geq 0}$ is an $(\mathcal{F}_t)$-martingale for each $k \geq 1$.

Note that by convention, the term *stochastic integral* stands for both a specific random variable $\omega \longmapsto \int_0^t H_s(\omega) \, dB_s(\omega)$ for a fixed t as well as the entire process. Extension of the definition of stochastic integrals on a fixed time interval $[0,T]$ to those on the positive half line $[0,\infty)$ is straightforward and hence omitted here. Note also that by replacing $[B,B]$ with $[M,M]$ for a continuous square-integrable martingale M, we can easily extend the above framework for integrals driven by B to those driven by M.

The Itô formula corresponds to the chain rule in ordinary calculus and is an indispensable tool in the study of stochastic differential equations. Here, the simplest form of the Itô formula is provided when the integrator is a Brownian motion B. Let $f \in C^2(\mathbb{R})$, then for each $t \geqslant 0$, a.s.,

$$f(B_t) - f(0) = \int_0^t f'(B_s)\,dB_s + \frac{1}{2}\int_0^t f''(B_s)\,ds, \qquad (2.3)$$

which is also written in shorthand (differential form) as

$$df(B_t) = f'(B_t)\,dB_t + \frac{1}{2}f''(B_t)\,dt.$$

For the proof and further details see, e.g. [Ikeda and Watanabe (1989), Karatzas and Shreve (1991)]. In Chapter 6 a time-changed generalization of the Itô formula will be proved.

2.4 FPK equations for stochastic processes driven by Brownian motion

2.4.1 FPK equation associated with Brownian motion

Let $p(t, y)$ denote the density function of a one-dimensional Brownian motion (B_t), which implies that

$$p(t, y) = (2\pi t)^{-1/2} e^{-y^2/(2t)} \qquad (2.4)$$

for $t > 0$ and $y \in \mathbb{R}$. It can be verified through a direct calculation that $p(t, y)$ satisfies the following equation, known as the diffusion (or heat) equation:

$$\frac{\partial p(t, y)}{\partial t} = \frac{1}{2}\frac{\partial^2 p(t, y)}{\partial y^2}, \qquad t > 0,\ y \in \mathbb{R}.$$

In fact, this is the Fokker–Planck or forward Kolmogorov equation associated with the Brownian motion. The same equation can be obtained through an application of the Itô formula (2.3). Namely, for an arbitrary $f \in C_c^\infty(\mathbb{R})$, an infinitely differentiable function on $\mathbb{R}$ with compact support, taking expectations on both sides of the Itô formula gives, for each fixed $t > 0$,

$$\mathbb{E}[f(B_t)] - f(0) = \mathbb{E}\left[\int_0^t f'(B_s)\,dB_s\right] + \frac{1}{2}\mathbb{E}\left[\int_0^t f''(B_s)\,ds\right].$$

Smoothness of f implies that $(f'(B_s))_{s \in [0,T]} \in \mathcal{L}^2[0, T]$ for a fixed $T > t$, and hence, the stochastic integral $\left(\int_0^t f'(B_s)\,dB_s\right)_{t \in [0,T]}$ is an $(\mathcal{F}_t)$-martingale, which has zero expectation. Hence, using the Fubini Theorem, the above equation becomes

$$\mathbb{E}[f(B_t)] - f(0) = \frac{1}{2}\int_0^t \mathbb{E}[f''(B_s)]\,ds.$$

Rewriting this with the density function $p(t, y)$, we obtain

$$\int_\mathbb{R} f(y)p(t, y)\,dy - f(0) = \frac{1}{2}\int_0^t \int_\mathbb{R} f''(y)p(s, y)\,dy\,ds$$

$$= \frac{1}{2}\int_0^t \int_\mathbb{R} f(y)\frac{\partial^2}{\partial y^2}p(s, y)\,dy\,ds,$$

where we used the integration-by-parts formula twice as well as the assumption that f has compact support. Taking the derivative with respect to t gives

$$\int_{\mathbb{R}} f(y) \frac{\partial}{\partial t} p(t, y) \, dy = \frac{1}{2} \int_{\mathbb{R}} f(y) \frac{\partial^2}{\partial y^2} p(t, y) \, dy.$$

Since $f \in C_c^\infty(\mathbb{R})$ is arbitrary, it follows that

$$\frac{\partial p(t, y)}{\partial t} = \frac{1}{2} \frac{\partial^2 p(t, y)}{\partial y^2}, \quad t > 0, \, y \in \mathbb{R}$$

with $p(0, y) = \delta_0(y)$, in the sense of generalized functions, where $\delta_0(y)$ is the Dirac delta function with mass at 0. Note that although the equation holds in the strict sense in the case of Brownian motion, such partial differential equations associated with general stochastic processes may be represented only in the sense of generalized functions. The above equation represents the Fokker–Planck or forward Kolmogorov equation associated with the Brownian motion and provides a macroscopic picture of behaviors of particles being predicted by the Brownian motion.

Associated with the Fokker–Planck equation is the backward Kolmogorov equation. Namely, if $\varphi \in C_c^2(\mathbb{R})$, then the function $u(t, x) = \mathbb{E}[\varphi(B_t)|B_0 = x]$ satisfies the following initial value problem, known as the Cauchy problem:

$$\frac{\partial u(t, x)}{\partial t} = \frac{1}{2} \frac{\partial^2 u(t, x)}{\partial x^2}, \quad t > 0, \, x \in \mathbb{R},$$

$$u(0, x) = \varphi(x).$$

As mentioned in Chapter 1, we refer to both the forward and backward Kolmogorov equations as the *Fokker–Planck–Kolmogorov (FPK)* equations.

The above derivation of the FPK equation in the case of Brownian motion suggests that the Itô formula provides a bridge between stochastic processes and associated partial differential equations satisfied by the transition probabilities. Applying this idea to stochastic processes determined as solutions of various types of stochastic differential equations and making macroscopic descriptions of those processes is one of the central themes of this book.

2.4.2   *FPK equations associated with SDEs driven by Brownian motion*

Now we demonstrate the derivation of the FPK equation associated with an Itô stochastic differential equation (SDE) driven by m-dimensional Brownian motion $(B_t)_{t \geq 0}$, where $B_t = (B_t^1, \ldots, B_t^m)$ with the components being independent one-dimensional Brownian motions. (Note that for convenience, we sometimes simply write X_t to denote an entire stochastic process $(X_t)_{t \geq 0}$.) The derivation of the FPK equation associated with an SDE requires the Itô formula for general d-dimensional Itô stochastic processes $Y_t = (Y_t^1, \ldots, Y_t^d)$ defined in the form

$$Y_t = Y_0 + \int_0^t b(s) \, ds + \int_0^t \sigma(s) \, dB_s, \tag{2.5}$$

where $Y_0 = (Y_0^1, \ldots, Y_0^d)$ is a d-dimensional random variable independent of Brownian motion B_t, and stochastic processes $b(t) \in \mathbb{R}^d$ and $\sigma(t) = (\sigma_{k\ell}(t); k = 1, \ldots, d, \ell = 1, \ldots, m) \in \mathbb{R}^{d \times m}$ are adapted to the filtration generated by B_t. The stochastic process Y_t in the componentwise form can be written as

$$Y_t^k = Y_0^k + \int_0^t b_k(s)ds + \sum_{\ell=1}^m \int_0^t \sigma_{k,\ell}(s)dB_s^\ell, \quad k = 1, \ldots, d. \tag{2.6}$$

Denoting

$$M_t = \int_0^t \sigma(s)dB_s \quad \text{and} \quad A_t = \int_0^t b(s)ds,$$

one can write Y_t in the form

$$Y_t = Y_0 + M_t + A_t. \tag{2.7}$$

It is not hard to verify that if $b(t) \in L^1[0, T]$ a.s., then A_t is a process of bounded variation. Moreover, if $\sigma(t) \in \mathcal{L}^2_{loc}[0, T]$, then as was noted in the previous section, the process M_t is a local martingale. Processes of the form (2.7) with a bounded variation process A_t and a local martingale M_t are called semimartingales. If $f = f(y_1, \ldots, y_d) \in C^2(\mathbb{R}^d)$, then for a d-dimensional continuous semimartingale Y_t, the following Itô formula is valid with probability one [Ikeda and Watanabe (1989), Karatzas and Shreve (1991)] (see also Section 5.2 of this book for the case of semimartingales that are not necessarily continuous):

$$f(Y_t) = f(Y_0)$$
$$+ \sum_{k=1}^d \int_0^t \frac{\partial f}{\partial y_k}(Y_s)dA_s^k + \sum_{k=1}^d \int_0^t \frac{\partial f}{\partial y_k}(Y_s)dM_s^k$$
$$+ \frac{1}{2} \sum_{j,k=1}^d \int_0^t \frac{\partial^2 f}{\partial y_j \partial y_k}(Y_s)d[M^j, M^k]_s.$$

Applying this formula to the Itô stochastic process of the form (2.5) for a function $f \in C^2(\mathbb{R}^d)$ yields

$$f(Y_t) = f(Y_0)$$
$$+ \int_0^t \left[\sum_{k=1}^d b_k(s)\frac{\partial f}{\partial y_k}(Y_s) + \sum_{j,k=1}^d a_{j,k}(s)\frac{\partial^2 f}{\partial y_j \partial y_k}(Y_s) \right] ds$$
$$+ \int_0^t \sum_{k=1}^d \sum_{\ell=1}^m \frac{\partial f}{\partial y_k}(Y_s)\sigma_{k,\ell}(Y_s)dB_s^\ell, \tag{2.8}$$

where $a_{j,k}(y)$, $j, k = 1, \ldots, d$, are entries of the matrix product $\frac{1}{2}\sigma(y)\sigma^T(y)$ ($\sigma^T(y)$ is the transpose of $\sigma(y)$), that is,

$$a_{j,k}(y) = \frac{1}{2} \sum_{\ell=1}^m \sigma_{k,\ell}(y)\sigma_{j,\ell}(y), \quad k, j = 1, \ldots d. \tag{2.9}$$

Formula (2.8) generalizes the Itô formula (2.3) for a one-dimensional Brownian motion to the case of general stochastic processes of the form (2.5). Now suppose a stochastic process X_t is determined as a solution to stochastic differential equation (SDE)

$$dX_t = b(X_t)dt + \sigma(X_t)dB_t, \quad X_0 = x, \tag{2.10}$$

where $x \in \mathbb{R}^d$ is non-random, the SDE is understood as

$$X_t = X_0 + \int_0^t b(X_s)ds + \int_0^t \sigma(X_s)dB_s, \tag{2.11}$$

and the mappings

$$b : \mathbb{R}^d \to \mathbb{R}^d, \quad \sigma : \mathbb{R}^d \to \mathbb{R}^{d \times m}$$

are Lipschitz continuous and satisfy the linear growth condition. Namely, there exist positive constants C_1 and C_2 such that for all $x, y \in \mathbb{R}^d$, the inequalities

$$\sum_{k=1}^d |b_k(x) - b_k(y)| + \sum_{k=1}^d \sum_{\ell=1}^m |\sigma_{k,\ell}(x) - \sigma_{k,\ell}(y)| \leqslant C_1 |x - y| \tag{2.12}$$

and

$$\sum_{k=1}^d |b_k(x)| + \sum_{k=1}^d \sum_{\ell=1}^m |\sigma_{k,\ell}(x)| \leqslant C_2(1 + |x|) \tag{2.13}$$

hold.

Applying the Itô formula to X_t in (2.11) with a test function $f \in C_c^\infty(\mathbb{R}^d)$ (i.e. f is a C^∞ function with compact support) yields

$$f(X_t) = f(X_0)$$
$$+ \int_0^t \left[\sum_{k=1}^d b_k(X_s) \frac{\partial f}{\partial y_k}(X_s) + \sum_{j,k=1}^d a_{j,k}(X_s) \frac{\partial^2 f}{\partial y_j \partial y_k}(X_s) \right] ds$$
$$+ \int_0^t \sum_{k=1}^d \sum_{\ell=1}^m \frac{\partial f}{\partial y_k}(X_s) \sigma_{k,\ell}(X_s) dB_s^\ell. \tag{2.14}$$

Further, let X_t^x be the stochastic process X_t in (2.11) conditioned on the event that $X_0 = x$, that is, $X_t^x = (X_t | X_0 = x)$. Denote by $u(t, y)$ the density function of X_t^x. Obviously $u(t, y)$ satisfies the condition $u(0, y) = \delta_x(y)$, where $\delta_x(y)$ is the Dirac delta function concentrated at x. Moreover, equation (2.14) can be rewritten as

$$f(X_t^x) = f(x)$$
$$+ \int_0^t \left[\sum_{k=1}^d b(X_s^x) \frac{\partial f}{\partial y_k}(X_s^x) + \sum_{j,k=1}^d a_{j,k}(X_s^x) \frac{\partial^2 f}{\partial y_j \partial y_k}(X_s^x) \right] ds$$
$$+ \int_0^t \sum_{k=1}^d \sum_{\ell=1}^m \frac{\partial f}{\partial y_k} \sigma_{k,\ell}(X_s^x) dB_s^\ell. \tag{2.15}$$

Taking the expectation on both sides of (2.15) and using the fact that the Itô integral driven by Brownian motion has zero expectation yields

$$\mathbb{E}[f(X_t^x)] = \int_{\mathbb{R}^d} f(y)u(t,y)dy$$

$$= f(x) + \int_0^t \int_{\mathbb{R}^d} \left[\sum_{k=1}^d b_k(y)\frac{\partial f}{\partial y_k}(y) + \sum_{j,k=1}^d a_{j,k}(y)\frac{\partial^2 f}{\partial y_j \partial y_k}(y) \right] u(s,y)\, dy\, ds.$$

$$(2.16)$$

Now differentiating equation (2.16) in the variable t, using the integration by parts, and noting that f has compact support, we obtain

$$\int_{\mathbb{R}^d} f(y)\frac{\partial u(t,y)}{\partial t}dy = \int_{\mathbb{R}^d} f(y)\left[-\sum_{k=1}^d \frac{\partial}{\partial y_k}\left(b_k(y)u(t,y)\right)\right.$$

$$\left. + \sum_{j,k=1}^d \frac{\partial^2}{\partial y_j \partial y_k}\left(a_{j,k}(y)u(t,y)\right)\right]dy, \quad t > 0.$$

Since f is an arbitrary test function, it follows that the density function $u(t,y)$ of the process X_t^x satisfies in the weak sense the Cauchy problem

$$\frac{\partial u(t,y)}{\partial t} = \mathcal{A}^* u(t,y), \quad t > 0, \ y \in \mathbb{R}^d, \tag{2.17}$$

$$u(0,y) = \delta_x(y), \quad y \in \mathbb{R}^d, \tag{2.18}$$

where the operator $\mathcal{A}^*$ is the second order elliptic differential operator defined by

$$\mathcal{A}^*\varphi(y) = -\sum_{k=1}^d \frac{\partial}{\partial y_k}\left(b_k(y)\varphi(y)\right) + \sum_{j,k=1}^d \frac{\partial^2}{\partial y_j \partial y_k}\left(a_{j,k}(y)\varphi(y)\right), \tag{2.19}$$

in which the positive definite matrix $(a_{j,k}(y); j,k = 1,\ldots,d)$ is defined in (2.9). Equation (2.17) represents the Fokker–Planck or forward Kolmogorov equation associated with the solution $X_t^x = (X_t|X_0 = x)$ of SDE (2.10) (or (2.11) with $X_0 = x$). The differential equation with the formally conjugate operator

$$\mathcal{A} = \sum_{k=1}^d b_k(x)\frac{\partial}{\partial x_k} + \sum_{j,k=1}^d a_{j,k}(y)\frac{\partial^2}{\partial x_j \partial x_k} \tag{2.20}$$

to the operator $\mathcal{A}^*$ defined in (2.19) represents the backward Kolmogorov equation

$$\frac{\partial u(t,x)}{\partial t} = \mathcal{A}u(t,x), \quad t > 0, \ x \in \mathbb{R}^d. \tag{2.21}$$

The backward and forward Kolmogorov equations can also be established for transition probabilities $P^X(t,x,B) = \mathbb{P}(X_t \in B|X_0 = x)$ of the stochastic process X_t, initially being at the position $x \in \mathbb{R}^d$ and arriving at the Borel set $B \subset \mathbb{R}^d$ at time t. If X_t is a solution to SDE (2.10), then $P^X(t,x,dy)$ satisfies in the weak sense the partial differential equations

$$\frac{\partial P^X(t,x,dy)}{\partial t} = \mathcal{A}\, P^X(t,x,dy), \tag{2.22}$$

where $\mathcal{A}$ acts on the backward variable x, and

$$\frac{\partial P^X(t, x, dy)}{\partial t} = \mathcal{A}^* P^X(t, x, dy), \tag{2.23}$$

where $\mathcal{A}^*$ acts on the forward variable y (see details in [Stroock (2003), Umarov (2015b)]). Equations (2.22) and (2.21), in which $\mathcal{A}$ acts on the backward variable x, are called *backward Kolmogorov equations*. Equations (2.23) and (2.17), where $\mathcal{A}^*$ acts on the forward variable y, are called *forward Kolmogorov equations*. The latter in the physics literature are called Fokker–Planck equations. Again, in this book, unifying the terminology, we call both equations the *Fokker–Planck–Kolmogorov (FPK) equations*.

2.4.3   *Connection with semigroup theory*

For our considerations, it is worth mentioning a connection of solutions of SDEs driven by Brownian motion with the semigroup theory. Note that the mapping

$$T_t f(x) = \mathbb{E}[f(X_t)|X_0 = x], \tag{2.24}$$

where X_t is a solution to SDE (2.10), defines a linear operator, for instance, in the Banach space $C_0(\mathbb{R}^d)$ of continuous functions vanishing at infinity equipped with the sup-norm. It follows from the definition that $T_0 f(x) = f(x)$, that is, $T_0 = I$, the identity operator. It also follows from the Markovian property of X_t that $T_s T_t f(x) = T_{t+s} f(x)$ (see, e.g. [Stroock (2003)]).

By definition (see [Engel and Nagel (1999)]), a one-parameter family $\{T_t; t \geqslant 0\}$ of linear operators defined on a Banach space $\mathcal{X}$ is called a *strongly continuous semigroup* if it satisfies the following conditions: (a) $T_0 = I$, the identity operator on $\mathcal{X}$; (b) $T_t T_s = T_{t+s}$ for all $t, s \geqslant 0$; and (c) $T_t \varphi \to T_{t_0} \varphi$ for all $\varphi \in \mathcal{X}$ in the norm of $\mathcal{X}$ as $t \to t_0$. Further, a linear operator A defined as $Af = \lim_{t\to 0+} t^{-1}(T_t f - f)$, provided the limit exists, is called the *infinitesimal generator* of the semigroup $\{T_t\}$. The domain of A is the set of elements $\varphi \in \mathcal{X}$ for which the above limit exists. The semigroup $\{T_t\}$ and its infinitesimal generator A are connected through the relation $T_t f(x) = e^{tA} f(x)$ where the operator e^{tA} is defined in an appropriate way.

Returning to the connection of the the solution X_t of SDE (2.10) with the semigroup $\{T_t\}$ defined by the family of linear operators in (2.24), notice that the operator $\mathcal{A}$ in (2.20) is the infinitesimal generator of the semigroup $\{T_t\}$ in (2.24). Moreover, the function $v(t, x) = T_t f(x) = \mathbb{E}[f(X_t)|X_0 = x]$ solves the Cauchy problem

$$\frac{\partial v(t, x)}{\partial t} = \mathcal{A} v(t, x), \quad t > 0, \ x \in \mathbb{R}^d,$$

$$v(0, x) = f(x), \quad x \in \mathbb{R}^d.$$

To illustrate this connection, suppose $m = d$, $b(x) = 0$, and the matrix $\sigma(x)$ has entries $\sigma_{k,\ell} = \delta_{k,\ell}$, $k, \ell = 1, \ldots, d$, where $\delta_{k,\ell}$ is the Kronecker symbol. In this case $X_t = B_t$, d-dimensional Brownian motion. Let $B_t^x = (B_t|B_0 = x)$. The condition

$B_0 = x$, that is $B_0^1 = x_1, \ldots, B_0^d = x_d$, and (2.4) imply that the density function of B_t^x is

$$f_{B_t^x}(y) = \prod_{k=1}^{d} (2\pi t)^{-1/2} e^{-(y_k - x_k)^2/(2t)}$$

$$= \frac{1}{(2\pi t)^{d/2}} e^{-\frac{|y-x|^2}{2t}}. \tag{2.25}$$

Assume $f \in C_0(\mathbb{R}^d)$. Then due to (2.25),

$$T_t f(x) = \mathbb{E}[f(B_t)|B_0 = x]$$

$$= \mathbb{E}\left[f(B_t^x)\right]$$

$$= \frac{1}{(2\pi t)^{d/2}} \int_{\mathbb{R}^d} e^{-\frac{|y-x|^2}{2t}} f(y) dy. \tag{2.26}$$

We will see in Chapter 4 that the latter can be written as $T_t f(x) = e^{\frac{1}{2}\Delta} f(x)$. Hence, the infinitesimal generator corresponding to Brownian motion B_t is the Laplace operator with the factor $1/2$. It is also well known that the function $v(t, x) = T_t f(x)$ defined in (2.26) solves the Cauchy problem

$$\frac{\partial v(t, x)}{\partial t} = \frac{1}{2} \Delta v(t, x), \quad t > 0, \ x \in \mathbb{R}^d,$$

$$v(0, x) = f(x), \quad x \in \mathbb{R}^d.$$

2.4.4 Markovian processes and the Chapman–Kolmogorov equation

This subsection introduces nonhomogeneous FPK equations, which are associated with Markovian processes that are nonhomogeneous in time. Readers whose main interests are in various fractional generalizations of homogeneous FPK equations of the form (2.17) and (2.21) can skip this subsection on a first reading.

A stochastic process X_t defined on a probability space $(\Omega, \mathcal{F}, \mathbb{P})$ with a filtration $(\mathcal{F}_t)$ is called *Markovian* with respect to $(\mathcal{F}_t)$ if for any $t \geq s$, Borel set $B \subset \mathbb{R}^d$, and $E \in \mathcal{F}_s$, one has $\mathbb{P}(X_t \in B|E, X_s = y) = \mathbb{P}(X_t \in B|X_s = y)$. That is, the future state of X_t does not depend on past events $E \in \mathcal{F}_s$, but it depends only on the current state $X_s = y$. In terms of transition probabilities $\mathbb{P}(t, x|s, y)$, $t \geq s$, this property takes the form: for any $s < \tau < t$,

$$\mathbb{P}(t, x|s, y) = \int \mathbb{P}(t, x|\tau, z) \mathbb{P}(\tau, z|s, y) dz. \tag{2.27}$$

This is called the *Chapman–Kolmogorov equation*. Notice that if the process is not Markovian, then this relationship would take the form

$$\mathbb{P}(t, x|s, y) = \int \mathbb{P}(t, x|\tau, z; s, y) \mathbb{P}(\tau, z|s, y) dz.$$

Suppose that the transition probabilities $\mathbb{P}(t, x|s, y)$ satisfy the conditions:

(i) $\lim\limits_{\tau \to 0} \mathbb{P}(t + \tau, x|t, y) = \mathbb{P}(t+, x|t, y);$

(ii) $\lim\limits_{\tau \to 0} \dfrac{1}{\tau} \int_{|x-y|<\varepsilon} (x_i - y_i)\mathbb{P}(t + \tau, x|t, y)dx = b_i(t, y) + O(\varepsilon), \quad \varepsilon \to 0;$

(iii) $\lim\limits_{\tau \to 0} \dfrac{1}{\tau} \int_{|x-y|<\varepsilon} (x_i - y_i)(x_j - y_j)\mathbb{P}(t + \tau, x|t, y)dx = a_{i,j}(t, y) + O(\varepsilon), \quad \varepsilon \to 0,$

uniformly in t, x, and y for all i and j. Under these conditions, using the Chapman–Kolmogorov equation (2.27), one can show that $\mathbb{P}(t, x|s, y)$ satisfies the partial differential equation

$$\frac{\partial \mathbb{P}(t, x|s, y)}{\partial t} = - \sum_{i=1}^{d} \frac{\partial}{\partial x_i}[b_i(t, x)\mathbb{P}(t, x|s, y)]$$

$$+ \frac{1}{2} \sum_{i,j=1}^{d} \frac{\partial^2}{\partial x_i \partial x_j}[a_{i,j}(t, x)\mathbb{P}(t, x|s, y)]$$

$$+ \int_{\mathbb{R}^d} [\mathbb{P}(t+, x|t, z)\mathbb{P}(t, z|s, y) - \mathbb{P}(t+, z|t, x)\mathbb{P}(t, x|s, y)]dz. \qquad (2.28)$$

The derivation of this equation is essentially the same as the derivation of the FPK equation presented above. One can show that the stochastic process X_t is continuous if and only if $\mathbb{P}(t+, x|t, y) = 0$ for all $t \geqslant 0$ and $x, y \in \mathbb{R}^d$ with $x \neq y$. In this case, equation (2.28) reduces to the forward FPK equation

$$\frac{\partial \mathbb{P}(t, x|s, y)}{\partial t} = - \sum_{i=1,d} \frac{\partial}{\partial x_i}[b_i(t, x)\mathbb{P}(t, x|s, y)]$$

$$+ \frac{1}{2} \sum_{i,j=1}^{d} \frac{\partial^2}{\partial x_i \partial x_j}[a_{i,j}(t, x)\mathbb{P}(t, x|s, y)]. \qquad (2.29)$$

Equation (2.29) represents the forward FPK equation in terms of transition probabilities with t-dependent coefficients $b_i(t, x)$ and $a_{i,j}(t, x)$, $i, j = 1, \ldots, d$, which generalizes equation (2.23). Dependence of coefficients of the FPK equation on the variable t means that the governing law changes not only from position to position, but also from time to time. Equation (2.29) is called a *nonhomogeneous* FPK equation, while FPK equations (2.23) and (2.22) are called *homogeneous*. Taking the adjoint of the operator on the right of (2.29), the reader can easily obtain the backward version of the nonhomogeneous FPK equation (2.29) as

$$\frac{\partial \mathbb{P}(t, x|s, y)}{\partial s} = \sum_{i=1}^{d} b_i(s, y)\frac{\partial \mathbb{P}(t, x|s, y)}{\partial y_i}$$

$$+ \frac{1}{2} \sum_{i,j=1}^{d} a_{i,j}(s, y)\frac{\partial^2 \mathbb{P}(t, x|s, y)}{\partial y_i \partial y_j}. \qquad (2.30)$$

Introducing a t-dependent differential operator

$$\mathcal{A}(t) = \sum_{k=1}^{d} b_k(t, x)\frac{\partial}{\partial x_k} + \sum_{j,k=1}^{d} a_{j,k}(t, x)\frac{\partial^2}{\partial x_j \partial x_k}, \qquad (2.31)$$

it is possible to write the nonhomogeneous forward and backward FPK equations in terms of the density function $u(t, x)$ of $X_t^x = (X_t | X_0 = x)$. Namely, the backward FPK equation in the nonhomogeneous case is

$$\frac{\partial u(t, x)}{\partial t} = \mathcal{A}(t) u(t, x), \quad t > 0, \ x \in \mathbb{R}^d,$$

and the forward FPK equation is

$$\frac{\partial u(t, x)}{\partial t} = \mathcal{A}^*(t) u(t, x), \quad t > 0, \ x \in \mathbb{R}^d, \tag{2.32}$$

where $\mathcal{A}^*(t)$ is the formally adjoint operator to $\mathcal{A}(t)$.

2.4.5 *FPK equations associated with SDEs driven by Brownian motion in bounded domains*

This subsection discusses FPK equations associated with SDEs defined in bounded domains, where many applications arise. Readers initially interested in stochastic processes defined on the whole space $\mathbb{R}^d$ can skip this subsection on a first reading.

If the solution process X_t of the Itô-type SDE with (t, x)-dependent drift and diffusion coefficients moves in a bounded region $\Omega \subset \mathbb{R}^d$ with a smooth boundary $\partial \Omega$, then for all $t \geq 0$, the probability $\mathbb{P}(X_t \in \mathbb{R}^d \backslash \Omega)$ of being out of the region Ω is zero. The associated FPK equation in this case needs to be supplemented by boundary conditions. In order to see in what form the boundary conditions are given, we need to introduce the notion of *probability current,* a d-dimensional vector field $\mathbf{\Phi}(t, x)$, components of which are defined as

$$\Phi_k(t, x) = b_k(t, x) p(t, x) - \sum_{j=1}^{d} \frac{\partial}{\partial x_j} \Big[a_{j,k}(t, x) p(t, x) \Big], \quad k = 1, \ldots, d.$$

Using the probability current, one can write the forward FPK equation (2.32) in the form of a conservation law:

$$\frac{\partial p(t, x)}{\partial t} + \sum_{k=1}^{d} \frac{\partial \Phi_k(t, x)}{\partial x_k} = 0,$$

or equivalently,

$$\frac{\partial p(t, x)}{\partial t} + \nabla \cdot \mathbf{\Phi}(t, x) = 0,$$

where $\nabla = (\frac{\partial}{\partial x_1}, \ldots, \frac{\partial}{\partial x_d})$ is the gradient operator and the symbol "$\cdot$" means the dot product of two vector objects.

Let S be a $(d-1)$-dimensional hyper-surface in Ω, and $\mathbf{n}_x$, $x \in S$, be an outward normal to S at the point $x \in S$. Then the total flow of probabilities through the hyper-surface S can be calculated by the surface integral

$$\int_S \mathbf{\Phi}(t, x) \cdot \mathbf{n}_x \, dS. \tag{2.33}$$

Two boundary conditions are common in the study of random processes in a bounded region:

(1) *reflecting* boundary condition, and
(2) *absorbing* boundary condition.

The reflecting boundary condition means that there is no probability flow across the boundary, and hence the surface integral in (2.33) with $S = \partial\Omega$ is zero. Therefore,

$$\mathbf{\Phi}(t, x) \cdot \mathbf{n}_x = 0, \quad x \in \partial\Omega. \tag{2.34}$$

In this case X_t will stay in the region Ω forever and be reflected when X_t reaches the boundary $\partial\Omega$. The absorbing boundary means that the probability for X_t being on the boundary is zero, that is $\mathbb{P}(X_t \in \partial\Omega) = 0$, leading to the condition

$$p(t, x) = 0, \quad x \in \partial\Omega. \tag{2.35}$$

In this case X_t will be absorbed by the boundary as soon as X_t reaches the boundary. More general cases of boundary conditions will be discussed in Chapter 4.

Let a boundary operator $\mathcal{B}$ be defined in the form

$$\mathcal{B}\varphi(x) = \mu(x)\varphi(x) + \nu(x)\frac{\partial\varphi}{\partial\mathbf{n}_x}, \quad x \in \partial\Omega, \tag{2.36}$$

where $\mathbf{n}_x$ is the outward normal at the point $x \in \partial\Omega$ and the functions $\mu(x)$ and $\nu(x)$ are continuous on the boundary $\partial\Omega$ and determined by the application. In order to define operators that will be used in the FPK equations associated with the SDEs in a bounded domain, we introduce the following space:

$$C_{\mathcal{B}}^2(\Omega) := \{\varphi \in C^2(\Omega) : \mathcal{B}\varphi(x) = 0, \, x \in \partial\Omega\}$$

with the boundary operator $\mathcal{B}$ defined in (2.36). For each $t > 0$, let $A_{\mathcal{B}}(t)$ denote the restriction of $A(t)$ in (2.31) to the space $C_{\mathcal{B}}^2(\Omega)$. In other words, the operator $A_{\mathcal{B}}(t)$ formally is the same as the operator $A(t)$ defined in (2.31), but with the domain $\mathrm{Dom}(A_{\mathcal{B}}(t)) = C_{\mathcal{B}}^2(\Omega)$. The operator $A_{\mathcal{B}}(t)$ is linear and maps the space $C_{\mathcal{B}}^2(\Omega)$ to $C(\Omega)$.

The boundary conditions for stochastic processes in a domain with an absorbing or reflecting boundary discussed above can be both reduced to the boundary condition

$$\mathcal{B}p(t, x) = 0, \quad t > 0, \, x \in \partial\Omega.$$

Therefore, the forward and backward FPK equations in the case of bounded domain in terms of the density function $u(t, x)$ of $X_t^x = (X_t|X_0 = x)$ can be formulated with the help of the operator $A_{\mathcal{B}}(t)$. Namely, the backward FPK equation is given by

$$\frac{\partial u(t, x)}{\partial t} = A_{\mathcal{B}}(t)u(t, x), \quad t > 0, \, x \in \Omega,$$

and the forward FPK equation is given by

$$\frac{\partial u(t, x)}{\partial t} = A_{\mathcal{B}}^*(t)u(t, x), \quad t > 0, \, x \in \Omega,$$

where $A_{\mathcal{B}}^*(t)$ is the formally adjoint operator to $A_{\mathcal{B}}(t)$. In Sections 4.6 and 7.7, we will consider a general case of the FPK equations on bounded domains and their fractional generalizations.

Chapter 3

Fractional Calculus

In this chapter we introduce and study properties of four different types of fractional order derivatives, namely the fractional derivatives in the senses of Riemann–Liouville, Caputo–Djrbashian, Liouville–Weyl and Riesz–Feller, and their distributed order generalizations. These derivatives are required in Chapter 7 to describe fractional Fokker–Planck–Kolmogorov equations. Fractional order derivatives in fact are defined as inverses of fractional order integration operators. Therefore this chapter starts with introducing the notion of fractional integral and studying some of its properties.

3.1 The Riemann–Liouville fractional derivative

Definition 3.1. Let a function $f(t)$ be defined and measurable on an interval (a, b), $a < b \leqslant \infty$. The *fractional integral* of order $\alpha > 0$ of the function f is defined by

$$_aJ_t^\alpha f(t) = \frac{1}{\Gamma(\alpha)} \int_a^t (t - \tau)^{\alpha-1} f(\tau) d\tau, \ t \in (a, b),$$

where $\Gamma(\alpha)$ is Euler's gamma function, that is,

$$\Gamma(\alpha) = \int_0^\infty t^{\alpha-1} e^{-t} dt.$$

If $\alpha = 0$, then we agree that $_aJ_t^0 = I$, the identity operator. If $\alpha = 1$, then $_aJ_t^1$ denotes the usual integral operator and is simply denoted $_aJ_t$. Also, in the case $a = 0$, we simply write J^α instead of $_0J_t^\alpha$.

Proposition 3.1. *For arbitrary $\alpha \geqslant 0$ and $\beta \geqslant 0$ the following semigroup property holds:*

$$_aJ_t^\alpha \, _aJ_t^\beta = _aJ_t^\beta \, _aJ_t^\alpha = _aJ_t^{\alpha+\beta}.$$

Proof. It suffices to show the equality $_aJ_t^\alpha \, _aJ_t^\beta = _aJ_t^{\alpha+\beta}$, which is obvious if either $\alpha = 0$, or $\beta = 0$. Assume that $\alpha > 0$ and $\beta > 0$. Then changing order of integration yields

$$_aJ_t^\alpha \, _aJ_t^\beta f(t) = \frac{1}{\Gamma(\alpha)} \int_a^t (t-\tau)^{\alpha-1} \left(\frac{1}{\Gamma(\beta)} \int_a^\tau (\tau-u)^{\beta-1} f(u) du \right) d\tau$$

$$= \frac{1}{\Gamma(\alpha)\Gamma(\beta)} \int_a^t f(u) \left(\int_u^t (t-\tau)^{\alpha-1} (\tau-u)^{\beta-1} d\tau \right) du.$$

The internal integral after substitution $\tau = t - (t-u)s$ takes the form

$$(t-u)^{\alpha+\beta-1} \int_0^1 s^{\alpha-1}(1-s)^{\beta-1} ds = B(\alpha,\beta)(t-u)^{\alpha+\beta-1},$$

where $B(\alpha,\beta)$ is Euler's beta function

$$B(\alpha,\beta) = \int_0^1 s^{\alpha-1}(1-s)^{\beta-1} ds.$$

The equality $_aJ_t^\alpha \, _aJ_t^\beta f(t) = \, _aJ_t^{\alpha+\beta} f(t)$ follows upon using the following relation between Euler's beta and gamma functions:

$$B(\alpha,\beta) = \frac{\Gamma(\alpha)\Gamma(\beta)}{\Gamma(\alpha+\beta)}, \tag{3.1}$$

thereby completing the proof. □

Definition 3.2. Let m be a positive integer and $m - 1 \leqslant \alpha < m$. The *fractional derivative* of order α of a function f *in the sense of Riemann–Liouville* is defined by

$$_aD_+^\alpha f(t) = \frac{1}{\Gamma(m-\alpha)} \frac{d^m}{dt^m} \int_a^t \frac{f(\tau)d\tau}{(t-\tau)^{\alpha+1-m}}, \tag{3.2}$$

provided the expression on the right exists. In the case $a = 0$, we simply write D_+^α instead of $_0D_+^\alpha$.

It is possible to write $_aD_+^\alpha$ in the operator form

$$_aD_+^\alpha = D^m \, _aJ_t^{m-\alpha}, \tag{3.3}$$

where we use the shorthand notation $D^m = \frac{d^m}{dt^m}$. This operator is the left-inverse to the fractional integration operator $_aJ_t^\alpha$. Indeed, due to Proposition 7.1,

$$_aD_+^\alpha \, _aJ_t^\alpha = D^m \, _aJ_t^{m-\alpha} \, _aJ_t^\alpha = D^m \, _aJ_t^m = I.$$

If $\alpha = m - 1$ in (3.3), then it follows from the equality $D^1 \, _aJ_t = I$ that $_aD_+^\alpha = D^m \, _aJ_t = D^{m-1}$. In this case the natural domain of $_aD_+^\alpha$ is $C^{m-1}(a,b)$, that is, $(m-1)$-times continuously differentiable functions defined on the interval (a,b). To explore the domain of $_aD_+^\alpha$ for any non-integer order α, consider the case $0 < \alpha < 1$. It follows from Definition 3.2 that if $0 < \alpha < 1$, then

$$_aD_+^\alpha f(t) = \frac{1}{\Gamma(1-\alpha)} \frac{d}{dt} \int_a^t \frac{f(\tau)d\tau}{(t-\tau)^\alpha}. \tag{3.4}$$

The operator form of $_aD_+^\alpha$ in this case is $_aD_+^\alpha = D^1 \, _aJ_t^{1-\alpha}$.

Let $C^\lambda[a, b]$ denote the class of Hölder continuous functions of order $\lambda > 0$ on an interval $[a, b]$. The following statement says that if f is Hölder continuous of order $\lambda \in (0, 1)$, then its fractional derivative of order $\alpha \in (0, \lambda)$ exists.

Proposition 3.2. [Samko et al. (1993)] *Let $f \in C^\lambda[a, b]$ with $\lambda \in (0, 1]$. Then for any $\alpha \in (0, \lambda)$, the fractional derivative ${}_aD_+^\alpha f(t)$ exists and can be represented in the form*

$$_aD_+^\alpha f(t) = \frac{f(a)}{\Gamma(1 - \alpha)(t - a)^\alpha} + \psi(t), \tag{3.5}$$

where $\psi \in C^{\lambda - \alpha}[a, b]$.

We note that in this proposition $C^1[a, b]$ corresponding to $\lambda = 1$ means the space of functions with continuous derivative on the interval $[a, b]$. The reader can verify that if $f(t) = (t - a)^\gamma$, where $\gamma > -1$, then

$$_aD_+^\alpha f(t) = \frac{\Gamma(1 + \gamma)}{\Gamma(1 + \gamma - \alpha)}(t - a)^{\gamma - \alpha}, \ t > a. \tag{3.6}$$

In particular, if $f(t) \equiv 1$ (i.e. $\gamma = 0$), then

$$_aD_+^\alpha f(t) = \frac{1}{\Gamma(1 - \alpha)(t - a)^\alpha}, \ t > a.$$

This shows that fractional derivatives of constant functions in the Riemann–Liouville sense are not zero, and $\psi(t) \equiv 0$ in representation (3.5).

3.2 The Caputo–Djrbashian fractional derivative

The operator form of the fractional derivative ${}_aD_*^\alpha$ of order α, $m - 1 \leqslant \alpha < m$, in the Caputo–Djrbashian sense is

$$_aD_*^\alpha = {}_aJ_t^{m-\alpha} D^m, \tag{3.7}$$

which is well defined, for instance, in the class of m-times continuously differentiable functions defined on an interval $[a, b)$ with $a < b$. Hence, we come to the following definition.

Definition 3.3. Let m be a positive integer and $m - 1 \leqslant \alpha < m$. The *fractional derivative* of order α of a function f *in the sense of Caputo–Djrbashian is defined by*

$$_aD_*^\alpha f(t) = \frac{1}{\Gamma(m - \alpha)} \int_a^t \frac{f^{(m)}(\tau)d\tau}{(t - \tau)^{\alpha + 1 - m}}, \ t > a, \tag{3.8}$$

provided the integral on the right exists. In the case $a = 0$, we simply write D_*^α instead of ${}_0D_*^\alpha$.

In contrast to the Riemann–Liouville fractional derivative, the Caputo–Djrbashian derivative of a constant function is zero, which immediately follows from the definition of ${}_aD_*^\alpha$. Moreover, since

$$
{}_aJ_t^\alpha\,{}_aD_*^\alpha f(t) = {}_aJ_t^\alpha\,{}_aJ_t^{m-\alpha}D^m f(t)
$$
$$
= {}_aJ_t^m D^m f(t)
$$
$$
= f(t) - \sum_{k=0}^{m-1} \frac{f^{(k)}(a)}{k!}(t-a)^k, \tag{3.9}
$$

the operator ${}_aD_*^\alpha$ is the right inverse to ${}_aJ_t^\alpha$ up to the additive polynomial

$$
- \sum_{k=0}^{m-1} \frac{f^{(k)}(a)}{k!}(t-a)^k.
$$

If the operator ${}_aD_*^\alpha$ is considered in the class of functions $f \in C^m[a,b]$ such that $f^{(k)}(a) = 0$ for $k = 0, \ldots, m-1$, then ${}_aJ_t^\alpha\,{}_aD_*^\alpha = I$, that is, ${}_aD_*^\alpha$ is the exact right inverse to ${}_aJ_t^\alpha$.

It follows from Definition 3.3 that if $0 < \alpha < 1$, then

$$
{}_aD_*^\alpha f(t) = \frac{1}{\Gamma(1-\alpha)} \int_a^t \frac{f'(\tau)d\tau}{(t-\tau)^\alpha}. \tag{3.10}
$$

The operator form of ${}_aD_*^\alpha$ in this case is ${}_aD_*^\alpha = {}_aJ_t^{1-\alpha}D^1$.

The Riemann–Liouville and Caputo–Djrbashian fractional derivatives are interrelated. Indeed, applying ${}_aD_+^\alpha$ to both sides of (3.9) and taking into account the fact that ${}_aD_+^\alpha$ is the left inverse to ${}_aJ_t^\alpha$ yields

$$
{}_aD_*^\alpha f(t) = {}_aD_+^\alpha \left(f(t) - \sum_{k=0}^{m-1} \frac{f^{(k)}(a)}{k!}(t-a)^k \right). \tag{3.11}
$$

Further, since (see (3.6))

$$
{}_aD_+^\alpha[(t-a)^k] = \frac{k!}{\Gamma(1+k-\alpha)}(t-a)^{k-\alpha},\ t > a,
$$

relationship (3.11) reduces to

$$
{}_aD_*^\alpha f(t) = {}_aD_+^\alpha f(t) - \sum_{k=0}^{m-1} \frac{f^{(k)}(a)}{\Gamma(1+k-\alpha)}(t-a)^{k-\alpha}. \tag{3.12}
$$

3.3   Laplace transform of fractional derivatives

Let $a = 0$ and recall that we simply write J^α, D_+^α and D_*^α respectively instead of ${}_0J_t^\alpha$, ${}_0D_+^\alpha$ and ${}_0D_*^\alpha$. Suppose f is a function defined on the positive real line $[0, \infty)$ such that $D_+^\alpha f(t)$ and $D_*^\alpha f(t)$ exist. Below we derive formulas for the Laplace transform of $D_+^\alpha f(t)$ and $D_*^\alpha f(t)$. By definition, the Laplace transform of a piecewise continuous function $f(t)$ defined on $[0, \infty)$ and satisfying the condition $|f(t)| \leqslant Me^{at}$ with some constants $M > 0$ and $a \geqslant 0$ is

$$
L[f](s) = \int_0^\infty f(t)e^{-st}dt, \quad s > a.
$$

Recall the first differentiation formula for the Laplace transform: for $m = 1, 2, \ldots$,

$$L[D^m f](s) = s^m L[f](s) - \sum_{k=0}^{m-1} D^k f(0) s^{m-1-k}. \qquad (3.13)$$

In Propositions 3.3–3.5, assume that the Laplace transform $L[f](s)$ of a function $f(t)$ exists for $s > a$ with $a \geqslant 0$.

Proposition 3.3. *Let $\alpha > 0$. Then the Laplace transform of $J^\alpha f(t)$ is*

$$L[J^\alpha f](s) = s^{-\alpha} L[f](s), \quad s > a. \qquad (3.14)$$

Proof. By definition,

$$L[J^\alpha f](s) = \frac{1}{\Gamma(\alpha)} \int_0^\infty e^{-st} \int_0^t (t - \tau)^{\alpha-1} f(\tau) \, d\tau \, dt. \qquad (3.15)$$

Changing the order of integration due to the Fubini theorem, the right hand side of (3.15) can be written as

$$\frac{1}{\Gamma(\alpha)} \int_0^\infty f(\tau) \int_\tau^\infty (t - \tau)^{\alpha-1} e^{-st} \, dt \, d\tau. \qquad (3.16)$$

The substitution $t - \tau = u/s$ in the internal integral reduces it into

$$\int_\tau^\infty (t - \tau)^{\alpha-1} e^{-st} \, dt = \Gamma(\alpha) \frac{e^{-s\tau}}{s^\alpha}. \qquad (3.17)$$

The latter and equations (3.15) and (3.16) imply (3.14). $\qquad \square$

Remark 3.1. One can derive useful implications from relationship (3.17). Indeed, evaluating both sides at $s = i\xi$, $\xi \neq 0$ and taking the limit as $\tau \to 0$ (in the sense of tempered distributions) yields

$$\int_0^\infty t^{\alpha-1} e^{-i\xi t} \, dt = \frac{\Gamma(\alpha)}{(i\xi)^\alpha}, \quad \xi \neq 0, \; \alpha > 0. \qquad (3.18)$$

Here $(i\xi)^\alpha$ is understood in the sense of the power of complex numbers. For the strict justification of (3.18), see e.g. [Umarov (2015b)].

Proposition 3.4. *Let m be a positive integer and $m-1 \leqslant \alpha < m$. Then the Laplace transform of $D_+^\alpha f(x)$ is*

$$L[D_+^\alpha f](s) = s^\alpha L[f](s) - \sum_{k=0}^{m-1} (D^k J^{m-\alpha} f)(0) s^{m-1-k}, \quad s > a. \qquad (3.19)$$

Proof. By the operator form of D_+^α in (3.3) and formulas (3.13) and (3.14),

$$L[D_+^\alpha f](s) = L[D^m J^{m-\alpha} f](s)$$

$$= s^m L[J^{m-\alpha} f](s) - \sum_{k=0}^{m-1} (D^k J^{m-\alpha} f)(0) s^{m-1-k}$$

$$= s^\alpha L[f](s) - \sum_{k=0}^{m-1} (D^k J^{m-\alpha} f)(0) s^{m-1-k},$$

which completes the proof. $\qquad \square$

Proposition 3.5. *Let m be a positive integer and $m-1 \leqslant \alpha < m$. The Laplace transform of the Caputo–Djrbashian derivative of a function $f \in C^m[0,\infty)$ is*

$$L[D_*^\alpha f](s) = s^\alpha L[f](s) - \sum_{k=0}^{m-1} f^{(k)}(0)s^{\alpha-1-k}, \quad s > a. \tag{3.20}$$

Proof. By the operator form of D_*^α in (3.7) and formulas (3.14) and (3.13),

$$L[D_*^\alpha f](s) = L[J^{m-\alpha}D^m f](s) = s^{-(m-\alpha)}L[D^m f](s)$$

$$= s^{-(m-\alpha)}\left(s^m L[f](s) - \sum_{k=0}^{m-1} f^{(k)}(0)s^{m-1-k}\right)$$

$$= s^\alpha L[f](s) - \sum_{k=0}^{m-1} f^{(k)}(0)s^{\alpha-1-k},$$

which completes the proof.   □

For $\alpha \in (0,1)$, formulas (3.19) and (3.20) respectively take the forms:

$$L[D_+^\alpha f](s) = s^\alpha L[f](s) - (J^{1-\alpha}f)(0), \tag{3.21}$$

$$L[D_*^\alpha f](s) = s^\alpha L[f](s) - f(0)s^{\alpha-1}. \tag{3.22}$$

3.4   Distributed order differential operators

For our future considerations, distributed fractional order differential operators play an important role. In this section we present important properties of such operators.

Let $T > 0$ be an arbitrary number and m be an arbitrary positive integer. Let $f \in C^m[0,T]$. Let μ be a finite measure defined on the interval $[0,\nu]$ with $\nu \in (m-1, m]$ such that the function $\alpha \to D_*^\alpha f(t)$ is μ-integrable for all $t \in [0,T]$, where D_*^α is the Caputo–Djrbashian fractional derivative of order α. The operator D_μ defined by

$$D_\mu f(t) = \int_0^\nu D_*^\alpha f(t)d\mu(\alpha), \quad 0 < t \leqslant T, \tag{3.23}$$

is called a *distributed fractional order differential operator with mixing measure μ*.

The above definition of the distributed order differential operator is based on the Caputo–Djrbashian fractional derivative. Similarly, one can introduce the distributed order differential operator based on the Riemann–Liouville fractional derivative, i.e.

$$_{RL}D_\mu f(t) = \int_0^\nu D_+^\alpha f(t)d\mu(\alpha), \quad 0 < t \leqslant T. \tag{3.24}$$

A connection between these two operators is established in Proposition 3.10.

Example 3.1.

(1) If $\mu = \delta_\beta$, i.e. Dirac's delta with mass on $\beta \in (0,\nu]$, then $D_\mu = D_*^\beta$.

(2) If $\mu = \sum_{j=1}^{J} a_j \delta_{\alpha_j}$, where $a_j \in \mathbb{R}$ and $\alpha_j \in (0, \nu]$ for $j = 1, ..., J$, then

$$D_\mu f(t) = \sum_{j=1}^{J} a_j D_*^{\alpha_j} f(t).$$

(3) If $d\mu(\alpha) = a(\alpha)d\alpha$, where $a \in C[0, \nu]$ is a positive function, then

$$D_\mu f(t) = \int_0^\nu a(\alpha) D_*^\alpha f(t) d\alpha.$$

For applications analyzed in this book, the case $\nu = 1$ will be sufficient. Therefore, *in the remainder of this section, we only consider the case* $\nu = 1$. For properties of distributed order differential operators in the general case, we refer the reader to the book [Umarov (2015b)].

In the theory of distributed order differential operators, the following two functions play an important role: *the kernel function*

$$\mathcal{K}_\mu(t) = \int_0^1 \frac{t^{-\alpha} d\mu(\alpha)}{\Gamma(1 - \alpha)}, \quad t > 0, \tag{3.25}$$

and *the spectral function*

$$\Phi_\mu(s) = \int_0^1 s^\alpha d\mu(\alpha), \quad Re(s) > 0. \tag{3.26}$$

Proposition 3.6. [Kochubei (2008)] *Let $d\mu(\beta) = a(\beta)d\beta$, where $a \in C^3[0, 1]$ and $a(1) \neq 0$. Then*

(1) $\mathcal{K}_\mu(t) = \frac{a(1)}{t(\log t)^2} + O(\frac{1}{t(\log t)^3})$, $t \to 0$,
(2) $\Phi_\mu(s) = \frac{a(1)}{\log s} + O(\frac{1}{(\log s)^2})$, $s \to \infty$, *and*
(3) *if $a \in C[0, 1]$ and $a(0) \neq 0$, then $\Phi_\mu(s) \sim \frac{a(0)}{s \log s}$, $s \to 0$.*

Another description of the kernel function $\mathcal{K}_\mu$ is due to Meerschaert and Scheffler [Meerschaert and Scheffler (2006)]. Denote by $RV_\infty(\gamma)$ the set of functions *regularly varying* at infinity with exponent γ, that is, the eventually positive functions with behavior $g(\lambda t)/g(t) \to \lambda^\gamma$ as $t \to \infty$ for any $\lambda > 0$. The set $RV_0(\gamma)$ of functions that are regularly varying at zero with exponent γ can be defined in an analogous manner. Functions regularly varying with exponent $\gamma = 0$ are said to be *slowly varying*.

Proposition 3.7. [Meerschaert and Scheffler (2006)] *Let $d\mu(\alpha) = a(\alpha)d\alpha$ in (3.25). If $a \in RV_0(\beta - 1)$ for some $\beta > 0$, then there exists $K^* \in RV_\infty(0)$ such that $\mathcal{K}_\mu(t) = (\log t)^{-\beta} K^*(\log t)$. Therefore, $\mathcal{K}_\mu(t) = M(\log t)$ for some $M \in RV_\infty(-\beta)$ and $\mathcal{K}_\mu \in RV_\infty(0)$; thus, $\mathcal{K}_\mu$ is slowly varying at infinity. Conversely, if $\mathcal{K}_\mu(t) = M(\log t)$ for some $M \in RV_\infty(-\beta)$ and $\beta > 0$, then $a \in RV_0(\beta - 1)$.*

Proposition 3.8. *The distributed order differential operator D_μ defined in (3.23) has the representation*

$$D_\mu f(t) = \left(\mathcal{K}_\mu * \frac{df}{dt}\right)(t), \tag{3.27}$$

where $*$ denotes the convolution operation and the kernel function $\mathcal{K}_\mu$ is defined by (3.25).

Proof. By the definition of the operator D_μ,

$$
\begin{aligned}
D_\mu f(t) &= \int_0^1 D_*^\alpha f(t)d\mu(\alpha) \\
&= \int_0^1 \frac{1}{\Gamma(1-\alpha)} \int_0^t \frac{\frac{df(\tau)}{d\tau}d\tau}{(t-\tau)^\alpha} d\mu(\alpha) \\
&= \int_0^t \left(\int_0^1 \frac{(t-\tau)^{-\alpha}}{\Gamma(1-\alpha)} d\mu(\alpha)\right) \frac{df(\tau)}{d\tau} d\tau
\end{aligned} \tag{3.28}
$$

Due to (3.25), the internal integral in equation (3.28) is precisely $\mathcal{K}_\mu(t-\tau)$. Hence, the convolution in (3.27) follows. □

A similar argument yields the following proposition.

Proposition 3.9. *The distributed fractional order operator $_{RL}D_\mu$ defined in (3.24) has the representation*

$$_{RL}D_\mu f(t) = \frac{d}{dt}(\mathcal{K}_\mu * f)(t).$$

The following proposition establishes a connection between distributed order differential operators based on the Caputo–Djrbashian and Riemann–Liouville derivatives.

Proposition 3.10. *The two distributed order differential operators D_μ and $_{RL}D_\mu$ are related to each other through the formula*

$$D_\mu f(t) = {}_{RL}D_\mu f(t) - f(0)\mathcal{K}_\mu(t), \quad t > 0, \tag{3.29}$$

where $\mathcal{K}_\mu$ is defined by (3.25).

Proof. It follows from equation (3.12) with $a = 0$ and $m = 1$ that

$$D_*^\alpha f(t) = D_+^\alpha f(t) - \frac{f(0)}{\Gamma(1-\alpha)}t^{-\alpha}.$$

Integrating both sides with respect to the measure μ yields (3.29). □

Proposition 3.11. *The Laplace transform of the kernel function $\mathcal{K}_\mu$ in (3.25) is given by*

$$L[\mathcal{K}_\mu](s) = \int_0^1 s^{\alpha-1}d\mu(\alpha) = \frac{\Phi_\mu(s)}{s}, \quad s > 0, \tag{3.30}$$

where Φ_μ is the spectral function defined in (3.26).

Proof. The Laplace transform of $\mathcal{K}_\mu(t)$ is

$$L[\mathcal{K}_\mu](s) = \int_0^\infty e^{-st} \left(\int_0^1 \frac{t^{-\alpha} d\mu(\alpha)}{\Gamma(1-\alpha)} \right) dt$$

$$= \int_0^1 L\left[\frac{t^{-\alpha}}{\Gamma(1-\alpha)} \right](s) \, d\mu(\alpha).$$

The result follows due to formula (3.17) upon letting $\tau \to 0$ and using the definition of $\Phi_\mu(s)$ given in (3.26). $\qquad\square$

Proposition 3.11 can be applied for calculating the Laplace transforms of $D_\mu f$ and $_{RL}D_\mu f$.

Proposition 3.12. *Suppose that the Laplace transform $L[f](s)$ of a function f exists for $s > a$ with $a \geqslant 0$. Then the Laplace transform of $D_\mu f$ is*

$$L[D_\mu f](s) = \Phi_\mu(s)L[f](s) - f(0)\frac{\Phi_\mu(s)}{s}, \quad s > a. \tag{3.31}$$

The Laplace transform of $_{RL}D_\mu f$ is

$$L[_{RL}D_\mu f](s) = \Phi_\mu(s)L[f](s) - \int_0^1 (J^{1-\alpha}f)(0)d\mu(\alpha), \quad s > a. \tag{3.32}$$

Proof. Formulas (3.31) and (3.32) follow immediately from formulas (3.22) and (3.21), respectively, upon integrating with respect to the measure μ. Note that (3.31) is also an immediate consequence of Propositions 3.8 and 3.11. $\qquad\square$

3.5 The Liouville–Weyl fractional derivatives and the Fourier transform

This section concerns the fractional derivatives in the sense of Liouville–Weyl and some of their properties, which are needed to discuss the fractional derivatives in the sense of Riesz–Feller in Sections 3.6 and 3.7.

The fractional derivatives in the sense of Liouville–Weyl are defined with the help of the fractional integrals $_{-\infty}J^\alpha$ and $_xJ^\alpha_\infty$ with terminal points $\pm\infty$, which are given by

$$_{-\infty}J^\alpha f(x) = \frac{1}{\Gamma(\alpha)} \int_{-\infty}^x (x-y)^{\alpha-1}f(y)dy$$

and

$$_xJ^\alpha_\infty f(x) = \frac{1}{\Gamma(\alpha)} \int_x^\infty (y-x)^{\alpha-1}f(y)dy.$$

Let m be a positive integer and $m - 1 \leqslant \alpha < m$. Then the *Liouville–Weyl forward and backward fractional derivatives* of order α are defined as

$$_{-\infty}D^\alpha = D^m \, _{-\infty}J^{m-\alpha} \tag{3.33}$$

and

$$_x D_\infty^\alpha = (-D)^m \, _x J_\infty^{m-\alpha}, \tag{3.34}$$

respectively. Consequently, the respective explicit forms of the Liouville–Weyl forward and backward fractional derivatives are

$$_{-\infty} D^\alpha f(x) = \frac{1}{\Gamma(m-\alpha)} \frac{d^m}{dx^m} \int_{-\infty}^x \frac{f(y)}{(x-y)^{\alpha-m+1}} dy \tag{3.35}$$

and

$$_x D_\infty^\alpha f(x) = \frac{(-1)^m}{\Gamma(m-\alpha)} \frac{d^m}{dx^m} \int_x^\infty \frac{f(y)}{(y-x)^{\alpha-m+1}} dy. \tag{3.36}$$

The Liouville–Weyl fractional derivatives for f are well-defined if the function f satisfies some differentiability and decay conditions. Namely,

(i) the Liouville–Weyl forward derivative $_{-\infty} D^\alpha f(x)$ exists on $(-\infty, a]$ if $f \in C^\lambda(-\infty, a]$ has the asymptotic behavior $|f(x)| \sim |x|^{-m+\alpha-\varepsilon}$, $x \to -\infty$, and

(ii) the Liouville–Weyl backward derivative $_x D_\infty^\alpha f(x)$ exists on $[b, \infty)$ if $f \in C^\lambda[b, \infty)$ has the asymptotic behavior $|f(x)| \sim |x|^{-m+\alpha-\varepsilon}$, $x \to +\infty$,

where C^λ, $\lambda > \alpha$, is the Hölder class, a and b are finite numbers, and ε is an arbitrary positive number.

Remark 3.2. The Liouville–Weyl fractional derivatives can be interpreted as pseudo-differential operators with singular symbols. See Chapter 4; for further details, see [Umarov (2015b)].

Example 3.2. We will show that

$$_{-\infty} D^\alpha e^{ax} = a^\alpha e^{ax}$$

for an arbitrary $a > 0$. By definition (3.33),

$$_{-\infty} D^\alpha e^{ax} = D^m \left(_{-\infty} J^{m-\alpha} e^{ax} \right).$$

Therefore, we have to calculate

$$_{-\infty} J^{m-\alpha} e^{ax} = \frac{1}{\Gamma(m-\alpha)} \int_{-\infty}^x \frac{e^{ay}}{(x-y)^{\alpha-m+1}} dy.$$

Substituting $t = a(x-y)$ gives $_{-\infty} J^{m-\alpha} e^{ax} = e^{\alpha-m} e^{ax}$. Now, differentiating yields the desired result. Similarly, it can be shown that

$$_x D_\infty^\alpha e^{-ax} = a^\alpha e^{-ax}.$$

Let F denote the Fourier transform of a function $f \in L^1(-\infty, \infty)$. Namely,

$$F[f](\xi) = \int_{-\infty}^\infty f(x) e^{ix\xi} dx.$$

In fact, one can extend F to the space of tempered distributions (see [Umarov (2015b)]). In the latter space, the inverse Fourier transform exists and is given by

$$f(x) = \frac{1}{2\pi} \int_{-\infty}^{\infty} F[f](\xi) e^{-ix\xi} d\xi.$$

Proposition 3.13. *The following formulas hold:*

(1) $F[_{-\infty}J^\alpha f](\xi) = \frac{1}{(-i\xi)^\alpha} F[f](\xi), \ \xi \neq 0,$

(2) $F[_xJ^\alpha_\infty f](\xi) = \frac{1}{(i\xi)^\alpha} F[f](\xi), \ \xi \neq 0.$

Proof. Changing the order of integration in

$$F[_{-\infty}J^\alpha f](\xi) = \frac{1}{\Gamma(\alpha)} \int_{-\infty}^{\infty} \left(\int_{-\infty}^{x} f(y)(x - y)^{\alpha-1} dy \right) e^{ix\xi} dx$$

and then using the substitution $x - y = t$,

$$F[_{-\infty}J^\alpha f](\xi) = \frac{F[f](\xi)}{\Gamma(\alpha)} \int_0^\infty t^{\alpha-1} e^{it\xi} dt.$$

The first statement of the proposition now follows from relationship (3.18). The second statement can be proved in a similar manner. □

Proposition 3.14. *For the Fourier transform of forward and backward Liouville–Weyl fractional derivatives, the following formulas hold:*

(1) $F[_{-\infty}D^\alpha f](\xi) = (-i\xi)^\alpha F[f](\xi), \ \xi \neq 0,$

(2) $F[_xD^\alpha_\infty f](\xi) = (i\xi)^\alpha F[f](\xi), \ \xi \neq 0.$

Proof. It is well known that $F[D^m f](\xi) = (-i\xi)^m F[f](\xi)$ (see e.g. [Umarov (2015b)]). Using this and Proposition 3.13, we have

$$F[_{-\infty}D^\alpha f](\xi) = F[D^m {}_{-\infty}J^{m-\alpha} f](\xi)$$

$$= (-i\xi)^m \frac{1}{(-i\xi)^{m-\alpha}} F[f](\xi)$$

$$= (-i\xi)^\alpha F[f](\xi).$$

Similarly,

$$F[_xD^\alpha_\infty f](\xi) = F[(-1)^m D^m {}_xJ^{m-\alpha}_\infty f](\xi)$$

$$= (-1)^m(-i\xi)^m \frac{1}{(i\xi)^{m-\alpha}} F[f](\xi)$$

$$= (i\xi)^\alpha F[f](\xi),$$

which completes the proof. □

3.6   The Riesz potential and the Riesz–Feller fractional derivative

The fractional derivative in the sense of Riesz–Feller is the inverse to the Riesz potential. Therefore, we first introduce the Riesz potential and study some of its properties. In the one dimensional case it is also connected with fractional integrals with terminal points $\pm\infty$ introduced in Section 3.5.

Let $0 < \alpha < 1$. The integral

$$\mathcal{R}_0^\alpha f(x) = C_\alpha \int_{\mathbb{R}} \frac{f(y)}{|x-y|^{1-\alpha}} dy \tag{3.37}$$

defined for $f \in L^1(\mathbb{R})$ is called the *Riesz potential*. Here the constant C_α is the normalizing constant depending on α and equals [Umarov (2015b)]

$$C_\alpha = \frac{1}{2\Gamma(\alpha)\cos\frac{\pi\alpha}{2}}.$$

The connection of $\mathcal{R}_0^\alpha f(x)$ with fractional integrals $_{-\infty}J^\alpha f(x)$ and $_x J_\infty^\alpha f(x)$ is given by

$$\mathcal{R}_0^\alpha f(x) = \frac{1}{2\cos\frac{\pi\alpha}{2}}\left(_{-\infty}J^\alpha f(x) + _x J_\infty^\alpha f(x)\right). \tag{3.38}$$

Indeed,

$$\frac{1}{2\cos\frac{\pi\alpha}{2}}\left(_{-\infty}J^\alpha f(x) + _x J_\infty^\alpha f(x)\right)$$

$$= \frac{1}{2\Gamma(\alpha)\cos\frac{\pi\alpha}{2}}\left(\int_{-\infty}^x \frac{f(y)dy}{(x-y)^{1-\alpha}} + \int_x^\infty \frac{f(y)dy}{(y-x)^{1-\alpha}}\right)$$

$$= \frac{1}{2\Gamma(\alpha)\cos\frac{\pi\alpha}{2}}\int_{-\infty}^\infty \frac{f(y)dy}{|x-y|^{1-\alpha}}$$

$$= \mathcal{R}_0^\alpha f(x).$$

Using (3.38) one can extend the definition of the Riesz potential for arbitrary $\alpha > 0$ with $\alpha \notin \{2k+1; k = 0, 1, \ldots\}$ by setting

$$\mathcal{R}_0^\alpha = \frac{1}{2\cos\frac{\pi\alpha}{2}}\left(_{-\infty}J^\alpha + _x J_\infty^\alpha\right). \tag{3.39}$$

Obviously, $\mathcal{R}_0^\alpha$ is undefined for the values $\alpha \in \{2k+1; k = 0, 1, \ldots\}$.

Now we introduce the Riesz–Feller fractional derivative as the inverse to the Riesz potential. The *Riesz–Feller fractional derivative* of order α (analogously to (3.39)) is

$$\mathcal{D}_0^\alpha f(x) = \frac{-1}{2\cos\frac{\pi\alpha}{2}}\left(_{-\infty}D^\alpha f(x) + _x D_\infty^\alpha f(x)\right), \quad 0 < \alpha \notin \{2k+1; k = 0, 1, \ldots\}, \tag{3.40}$$

where $_{-\infty}D^\alpha$ and $_x D_\infty^\alpha$ are the Liouville–Weyl forward and backward fractional derivatives of order α defined in (3.35) and (3.36), respectively. Two important facts related to the Riesz–Feller fractional derivative are given below.

Proposition 3.15.

(1) The Fourier transform of the Riesz–Feller fractional derivative of f is
$$F[\mathcal{D}_0^\alpha f](\xi) = -|\xi|^\alpha F[f](\xi).$$

(2) Let $0 < \alpha < 2$. Then the Riesz–Feller fractional derivative of order α of a function f in the Hölder class $C^\lambda(\mathbb{R}) \cap L_\infty$, $\lambda > \alpha$, has the representation

$$\mathcal{D}_0^\alpha f(x) = \omega_\alpha \int_0^\infty \frac{\Delta_h^2 f(x)}{h^{1+\alpha}} dh, \qquad (3.41)$$

where $\Delta_h^2 f(x) = f(x+h) - 2f(x) + f(x-h)$ and $\omega_\alpha = (1/\pi)\Gamma(\alpha+1)\sin\frac{\pi\alpha}{2}$.

Proof. (1) Due to Proposition 3.14,

$$F[\mathcal{D}_0^\alpha f](\xi) = -\frac{1}{2\cos\frac{\pi\alpha}{2}}\left(F[_{-\infty}D^\alpha f](\xi) + F[_xD_\infty^\alpha f](\xi)\right)$$

$$= -\frac{1}{2\cos\frac{\pi\alpha}{2}}\left((-i\xi)^\alpha + (i\xi)^\alpha\right)F[f](\xi)$$

$$= -\frac{Re(i\xi)^\alpha F[f](\xi)}{\cos\frac{\pi\alpha}{2}}$$

$$= -\frac{|\xi|^\alpha\cos\frac{\pi\alpha}{2}F[f](\xi)}{\cos\frac{\pi\alpha}{2}} = -|\xi|^\alpha F[f](\xi).$$

(2) Here we prove (3.41) in the case $0 < \alpha < 1$. The general case will be proved in Theorem 3.3 in the multi-dimensional case using a different method. By the definition of the Riesz–Feller fractional derivative,

$$\mathcal{D}_0^\alpha f(x) = \frac{-1}{2\cos\frac{\pi\alpha}{2}}(_{-\infty}D^\alpha + {}_xD_\infty^\alpha)f(x)$$

$$= \frac{-D}{2\cos\frac{\pi\alpha}{2}}(_{-\infty}J^{1-\alpha} + {}_xJ_\infty^{1-\alpha})f(x)$$

$$= \frac{-D}{2\Gamma(1-\alpha)\cos\frac{\pi\alpha}{2}}\left(\int_{-\infty}^x \frac{f(y)dy}{(x-y)^\alpha} - \int_x^\infty \frac{f(y)dy}{(y-x)^\alpha}\right). \qquad (3.42)$$

The substitutions $x - y = u$ and $y - x = v$ in the two integrals on the right hand side of (3.42) give

$$\mathcal{D}_0^\alpha f(x) = \frac{-D}{2\Gamma(1-\alpha)\cos\frac{\pi\alpha}{2}}\left(\int_0^\infty \frac{f(x-u)du}{u^\alpha} - \int_0^\infty \frac{f(x+u)du}{u^\alpha}\right).$$

Further, using the equality $u^{-\alpha} = \alpha\int_u^\infty \frac{1}{y^{\alpha+1}}dy$ and changing the order of integration, we have

$$\frac{-\alpha}{2\Gamma(1-\alpha)\cos\frac{\pi\alpha}{2}}\left(\int_0^\infty f'(x-u)\int_u^\infty \frac{1}{y^{\alpha+1}}dy\,du - \int_0^\infty f'(x+y)\int_u^\infty \frac{1}{y^{\alpha+1}}dy\,du\right)$$

$$= \frac{\alpha}{2\Gamma(1-\alpha)\cos\frac{\pi\alpha}{2}}\int_0^\infty \frac{f(x-y) - 2f(x) + f(x+y)}{y^{1+\alpha}}dy.$$

This immediately yields (3.41) if one notices that the property

$$\Gamma(1-\alpha)\Gamma(1+\alpha) = \frac{\pi\alpha}{\sin\pi\alpha}$$

of Euler's gamma function implies

$$\frac{\alpha}{2\Gamma(1-\alpha)\cos\frac{\pi\alpha}{2}} = \frac{\Gamma(\alpha+1)}{\pi}\sin\frac{\pi\alpha}{2} = \omega_\alpha,$$

thereby completing the proof.   □

Corollary 3.1. *For the Fourier transforms of* $\mathcal{R}_0^\alpha f(x)$ *and* $\mathcal{D}_0^\alpha f(x)$, *the following formulas hold:*

(1) $F[\mathcal{R}_0^\alpha f](\xi) = |\xi|^{-\alpha} F[f](\xi);$
(2) $F[\mathcal{D}_0^\alpha f](\xi) = -|\xi|^\alpha F[f](\xi).$

Part (2) of this corollary implies that the α values in the operator $\mathcal{D}_0^\alpha f(x)$ can naturally be extended to $\alpha = 2$ as $\mathcal{D}_0^2 = \frac{d^2}{dx^2}$.

3.7   Multi-dimensional Riesz potentials and hyper-singular integrals

The n-dimensional *Riesz potential* is defined by

$$\mathbb{R}_0^\alpha f(x) = C_{n,\alpha} \int_{\mathbb{R}^n} \frac{f(y)\,dy}{|x-y|^{n-\alpha}}, \quad 0 < \alpha \notin \{2k+n; \, k = 0,1,\ldots\}, \tag{3.43}$$

where

$$C_{n,\alpha} = \frac{\Gamma(\frac{n-\alpha}{2})}{2^\alpha \pi^{n/2} \Gamma(\frac{\alpha}{2})}.$$

We denote by $\mathbb{D}_0^\alpha$ the inverse Riesz potential with "−" sign. That is,

$$\mathbb{D}_0^\alpha = -(\mathbb{R}_0^\alpha)^{-1}.$$

For our further considerations, the case $0 < \alpha < 2$ is important. In this case the operator $\mathbb{D}_0^\alpha$ can be represented as a hyper-singular integral

$$\mathbb{D}_0^\alpha f(x) = B_{n,\alpha} \int_{\mathbb{R}^n} \frac{f(x-y) - 2f(x) + f(x+y)}{|y|^{n+\alpha}}\,dy, \quad 0 < \alpha < 2, \tag{3.44}$$

where

$$B_{n,\alpha} = \frac{\alpha \Gamma(\frac{\alpha}{2})\Gamma(\frac{n+\alpha}{2})\sin\frac{\pi\alpha}{2}}{2^{2-\alpha}\pi^{1+n/2}}. \tag{3.45}$$

This fact follows from Theorem 3.3 proved below. Note that in the definition of $\mathbb{D}_0^\alpha$, the value $\alpha = 2$ is degenerate; in other words, if $\alpha = 2$, then $B_{n,\alpha} = 0$ and $\mathbb{D}_0^\alpha f(x) = 0$.

Recall that the Fourier transform of a function $f \in L^1(\mathbb{R}^n)$ is defined by

$$F[f](\xi) = \int_{\mathbb{R}^n} f(x)e^{i(x,\xi)}\,dx, \quad \xi \in \mathbb{R}^n,$$

where $(\cdot,\cdot)$ denotes the inner product in $\mathbb{R}^n$. The following statement is needed to establish representation (3.44).

Theorem 3.2. *Let $0 < \alpha \notin \{2k + n;\ k = 0, 1, \ldots\}$. Then for the Fourier transform of the n-dimensional Riesz potential, the formula*

$$F[\mathbb{R}_0^\alpha f](\xi) = \frac{1}{|\xi|^\alpha} F[f](\xi), \quad \xi \in \mathbb{R}^n,\ \xi \neq 0, \tag{3.46}$$

holds.

Proof. The Riesz potential can be expressed as a convolution. Namely,

$$\mathbb{R}_0^\alpha f(x) = \left(\frac{C_{n,\alpha}}{|x|^{n-\alpha}} * f \right)(x).$$

Applying the Fourier transform yields

$$F[\mathbb{R}_0^\alpha f](\xi) = \sigma_{\mathbb{R}_0^\alpha}(\xi) \cdot F[f](\xi),$$

where

$$\sigma_{\mathbb{R}_0^\alpha}(\xi) = C_{n,\alpha} F\left[\frac{1}{|x|^{n-\alpha}} \right](\xi).$$

Now we apply the following formula (see [Umarov (2015b)])

$$F\left[\frac{1}{|x|^\sigma} \right](\xi) = \frac{2^{n-\sigma} \pi^{n/2} \Gamma(\frac{n-\sigma}{2})}{\Gamma(\frac{\sigma}{2})} \frac{1}{|\xi|^{n-\sigma}}, \quad \xi \neq 0, \tag{3.47}$$

which is valid for all $\sigma \in \mathbb{C}$ such that $\sigma \neq -2m$ and $\sigma \neq n + 2m$ for any positive integer m. Setting $\sigma = n - \alpha$ in this formula yields

$$F\left[\frac{1}{|x|^{n-\alpha}} \right](\xi) = b_{\alpha,n} |\xi|^{-\alpha},$$

where

$$b_{\alpha,n} = \frac{2^\alpha \pi^{n/2} \Gamma(\frac{\alpha}{2})}{\Gamma(\frac{n-\alpha}{2})}.$$

Since $C_{n,\alpha} b_{\alpha,n} = 1$, formula (3.46) follows. $\square$

Since the operator $\mathbb{D}_0^\alpha$ is the inverse to $\mathbb{R}_0^\alpha$ with the minus sign, for any function f in the domain of $\mathbb{R}_0^\alpha$,

$$F[\mathbb{D}_0^\alpha \mathbb{R}_0^\alpha f](\xi) = -F[f](\xi) = \sigma_\alpha(\xi) \frac{1}{|\xi|^\alpha} F[f](\xi),$$

where $\sigma_\alpha(\xi) = -|\xi|^\alpha$. The function $\sigma_\alpha(\xi)$ is called the symbol of $\mathbb{D}_0^\alpha$ (see Chapter 4). Therefore, in order to establish equation (3.44), one needs to prove the following statement.

Theorem 3.3. *Let $0 < \alpha < 2$. Then the operator A defined by*

$$Af(x) = B_{n,\alpha} \int_{\mathbb{R}^n} \frac{f(x - y) - 2f(x) + f(x + y)}{|y|^{n+\alpha}} dy$$

has the symbol $\sigma_A(\xi) = -|\xi|^\alpha$.

Proof. It is known (see, e.g. [Hörmander (1983)]) that an operator A and its symbol $\sigma_A(\xi)$ are connected through $\sigma_A(\xi) = e^{i(x,\xi)} A e^{-i(x,\xi)}$, and hence, $A e^{-i(x,\xi)} = \sigma_A(\xi) e^{-i(x,\xi)}$. This implies that

$$\sigma_A(\xi) = \left(A e^{-i(x,\xi)} \right)\Big|_{x=0},$$

or equivalently,

$$\sigma_A(\xi) = B_{n,\alpha} \int_{\mathbb{R}^n} \frac{\Delta_y^2 e^{-i(x,\xi)}}{|y|^{n+\alpha}} dy \Big|_{x=0},$$

where $\Delta_y^2 f(x) = f(x+y) - 2f(x) + f(x-y)$. Therefore, we only need to verify that

$$\int_{\mathbb{R}^n} \frac{\Delta_y^2 e^{-i(x,\xi)}}{|y|^{n+\alpha}} dy \Big|_{x=0} = -\frac{|\xi|^\alpha}{B_{n,\alpha}}. \qquad (3.48)$$

The left hand side of (3.48) equals

$$\int_{\mathbb{R}^n} \frac{\Delta_y^2 e^{-i(x,\xi)}}{|y|^{n+\alpha}} dy \Big|_{x=0} = \int_{\mathbb{R}^n} \frac{e^{-i(x+y,\xi)} - 2 + e^{-i(x-y,\xi)}}{|y|^{n+\alpha}} dy \Big|_{x=0}$$

$$= \int_{\mathbb{R}^n} \frac{e^{-i(y,\xi)} - 2 + e^{i(y,\xi)}}{|y|^{n+\alpha}} dy$$

$$= F\left[\frac{1}{|y|^{n+\alpha}} \right](-\xi) - 2F\left[\frac{1}{|y|^{n+\alpha}} \right](0) + F\left[\frac{1}{|y|^{n+\alpha}} \right](\xi)$$

$$= \Delta_\xi^2 F\left[\frac{1}{|y|^{n+\alpha}} \right](0).$$

Taking $\sigma = n + \alpha$ in formula (3.47),

$$F\left[\frac{1}{|x|^{n+\alpha}} \right](\xi) = \frac{2^{-\alpha} \pi^{n/2} \Gamma(-\frac{\alpha}{2})}{\Gamma(\frac{n+\alpha}{2})} |\xi|^\alpha,$$

where $\Gamma(-\frac{\alpha}{2})$ is the value at $z = -\alpha/2 \in (-1,0)$ of the analytic continuation of Euler's gamma function $\Gamma(z)$ to the interval $(-1,0)$. It follows from this equality that

$$\Delta_\xi^2 F\left[\frac{1}{|y|^{n+\alpha}} \right](0) = \frac{2^{1-\alpha} \pi^{n/2} \Gamma(-\frac{\alpha}{2})}{\Gamma(\frac{n+\alpha}{2})} |\xi|^\alpha. \qquad (3.49)$$

Finally, using the relationship (see [Abramowitz and Stegun (1964)], formula 6.1.17)

$$-\frac{\alpha}{2} \Gamma\left(-\frac{\alpha}{2}\right) \Gamma\left(\frac{\alpha}{2}\right) = \frac{\pi}{\sin \frac{\pi\alpha}{2}}$$

in equation (3.49) yields (3.48).

$\square$

Remark 3.3. The operator $\mathbb{D}_0^\alpha$ in (3.44) is defined for $0 < \alpha < 2$. As was noted above, the value $\alpha = 2$ degenerate in this definition. However, the symbol of this operator is meaningful for $\alpha = 2$ as well: $\lim_{\alpha \to 2} \sigma_\alpha(\xi) = -|\xi|^2$. Since this is known to be the symbol of the Laplace operator Δ, it is natural to extend $\mathbb{D}_0^\alpha$ to $\alpha = 2$ by setting $\mathbb{D}_0^2 = \Delta$. It is also known that the operator $\mathbb{D}_0^\alpha$ is associated with the so-called spherically symmetric α-stable Lévy process (see Section 5.3; also see Example 4.1 in Section 4.3).

Chapter 4

Pseudo-differential operators associated with Lévy processes

Introduction

In this chapter a brief theory of pseudo-differential operators related to stochastic processes discussed in this book is presented. The theory of pseudo-differential operators in its modern form was developed in the 1960s by J. J. Kohn and L. Nirenberg [Kohn and Nirenberg (1965)] and by L. Hörmander [Hörmander (1965)]. In this now classical theory, symbols of operators are required to be infinitely differentiable on the cotangent bundle. However, symbols of pseudo-differential operators connected with Lévy or Lévy-type processes are not smooth. In Sections 4.3–4.6 we will discuss properties of the pseudo-differential operators associated with SDEs driven by Lévy and Lévy-type processes. In particular, some specific examples of pseudo-differential operators connected with Lévy processes are provided in Section 4.3.

4.1 Pseudo-differential operators

Let $\Omega \subset \mathbb{R}^d$ be a domain. L. Hörmander [Hörmander (1965)] constructed an algebra of pseudo-differential operators $OPS^m(\Omega)$ with smooth symbols $a(x, \xi) \in C^\infty(\Omega, T^*(\Omega))$, satisfying for any compact subset $K \subset \Omega$ the condition

$$|D_x^\beta D_\xi^\alpha a(x, \xi)| \leqslant C(1 + |\xi|)^{m-|\alpha|}, \ x \in K, \ \xi \in \mathbb{R}^d, \qquad (4.1)$$

for all multi-indices $\alpha = (\alpha_1, \dots, \alpha_d)$ and $\beta = (\beta_1, \dots, \beta_d)$. Here $T^*(\Omega)$ is the cotangent bundle of Ω, $|\alpha| = \alpha_1 + \cdots + \alpha_d$ is the length of α, $D_x = (-i\partial/\partial x_1, \dots, -i\partial/\partial x_d)$, $D_\xi = (-i\partial/\partial \xi_1, \dots, -i\partial/\partial \xi_d)$, and $C = C(\alpha, \beta, K)$ is a positive constant. By definition, $A \in OPS^m(\Omega)$ has the symbol $a(x, \xi) \in S^m(\Omega)$ if

$$Af(x) = \frac{1}{(2\pi)^d} \int_{\mathbb{R}^d} a(x, \xi) F[f](\xi) e^{-i(x, \xi)} d\xi, \ x \in \Omega, \qquad (4.2)$$

where $F[f](\xi)$ is the Fourier transform of f:

$$F[f](\xi) = \int_{\mathbb{R}^d} f(x) e^{i(x, \xi)} dx, \ \xi \in \mathbb{R}^d.$$

The operator in (4.2) is well defined on functions in the class $\mathcal{G}$, which contains infinitely differentiable functions f satisfying the estimate

$$(1 + |x|^2)^m |D^\alpha f(x)| \leqslant C_{m,\alpha}, \quad x \in \mathbb{R}^d,$$

for all $m = 0, 1, \ldots,$ and multi-indices α, with some constants $C_{m,\alpha} \geqslant 0$. The dual of $\mathcal{G}$ endowed with weak convergence is the space of tempered distributions and denoted by $\mathcal{G}'$.

A differential operator

$$A(x, D) = \sum_{|\alpha| \leqslant m} a_\alpha(x) D^\alpha$$

with coefficients $a_\alpha \in C^\infty(\Omega)$ is an example of an operator in $OPS^m(\Omega)$. The corresponding symbol is a polynomial in the variable ξ,

$$a(x, \xi) = \sum_{|\alpha| \leqslant m} a_\alpha(x) \xi^\alpha.$$

Thus, the algebra $OPS^m(\Omega)$ contains all the differential operators of order m with coefficients that are infinitely differentiable in Ω. In the algebra $OPS^m(\Omega)$, the addition and multiplication (composition) operations are well defined, as well as the adjoint operator. The reader is referred to books [Taylor (1981), Hörmander (1983), Jacob (2001, 2002, 2005), Umarov (2015b)] for details. The algebra $OPS^m(\Omega)$ contains the parametrices of all the elliptic operators of order m. A differential operator $A(x, D)$ with the symbol $a(x, \xi)$ is called *elliptic* if its main symbol

$$a_m(x, \xi) = \sum_{|\alpha| = m} a_\alpha(x) \xi^\alpha$$

satisfies the condition

$$a_m(x, \xi) \geqslant C_0 |\xi|^m$$

for all $x \in \Omega$ and $\xi \in \mathbb{R}^d$.

Though within $OPS^m(\Omega)$ one can describe parametrices of elliptic operators, the class $OPS^m(\Omega)$ is too restrictive to describe the so-called hypoellipticity of (pseudo) differential operators. If for arbitrary $f \in C^\infty(\Omega)$, a solution u of the equation $Au = f$ also is in $C^\infty(\Omega)$, then A is called *hypoelliptic*. Any elliptic pseudo-differential operator is hypoelliptic. There are hypoelliptic operators which are not elliptic. For example, the heat operator $\frac{\partial}{\partial t} - \Delta$ is not elliptic but hypoelliptic. The hypoellipticity of differential operators was described in works by Hörmander [Hörmander (1961), Hörmander (1967)], Egorov [Egorov (1967)], etc. The class of symbols $S^m_{\rho,\delta}(\Omega)$ depending on parameters $\rho \in (0, 1]$ and $\delta \in [0, 1]$ was introduced by Hörmander [Hörmander (1967)].

By definition, a symbol $a(x, \xi) \in C^\infty(\Omega \times \mathbb{R}^d \backslash \{0\})$ belongs to the class $S^m_{\rho,\delta}(\Omega)$ if $a(x, \xi)$ satisfies the condition

$$|D^\beta_x D^\alpha_\xi a(x, \xi)| \leqslant C(1 + |\xi|)^{m - \rho|\alpha| + \delta|\beta|}, \quad x \in K, \ \xi \in \mathbb{R}^d, \tag{4.3}$$

for all multi-indices α and β and compact sets $K \subset \Omega$. The corresponding class of pseudo-differential operators $OPS_{\rho,\delta}^m(\Omega)$ is wider than $OPS^m(\Omega)$ and within this class one can describe the hypoellipticity property of (pseudo) differential operators. The class of operators $OPS_{\rho,\delta}^m(\Omega)$ coincides with $OPS^m(\Omega)$ if $\rho = 1$, $\delta = 0$. We write S^m, $S_{\rho,\delta}^m$, OPS^m, and $OPS_{\rho,\delta}^m$, if $\Omega = \mathbb{R}^d$.

The proposition below reflects mapping properties of pseudo-differential operators with symbols from Hörmander classes. To formulate this proposition besides $\mathcal{G}$ and $\mathcal{G}'$, we need the following spaces. Denote by $\mathcal{D}(\Omega)$ the space of infinitely differentiable functions with compact support in Ω. The convergence $f_n \to f_0$ in $\mathcal{D}(\Omega)$ means uniform convergence $D^\alpha f_n \to D^\alpha f$ for all multi-indices α. The dual of $\mathcal{D}(\Omega)$ endowed with the weak convergence is the space of Schwartz distributions and denoted by $\mathcal{D}'(\Omega)$. The space of test functions $\mathcal{E}(\Omega)$ consists of infinitely differentiable functions with locally uniform convergence. Its dual $\mathcal{E}'(\Omega)$ is the space of distributions with compact support. For $s \in \mathbb{R}$, the Sobolev space $H^s(\mathbb{R}^d)$ is defined as

$$H^s(\mathbb{R}^d) = \{f \in \mathcal{G}' : (1 + |\xi|^2)^{s/2} F[f](\xi) \in L^2(\mathbb{R}^d)\}.$$

This is a Hilbert space with the inner product

$$(f, g) = \int_{\mathbb{R}^d} (1 + |\xi|^2)^s F[f](\xi) F[g](\xi) d\xi.$$

The space of Bessel potentials or the generalized Sobolev spaces $H_p^s(\mathbb{R}^d)$, $s \in \mathbb{R}$, $1 \leqslant p \leqslant \infty$, are defined by

$$H_p^s(\mathbb{R}^d) = \{f \in \mathcal{G}' : (1 + |\xi|^2)^{s/2} F[f](\xi) \in L^p(\mathbb{R}^d)\}.$$

This space is a Banach space with the norm $\|f\|_{p,s} = \|(1 + |\xi|^2)^{s/2} F[f]\|_{L^p(\mathbb{R}^d)}$. The Besov spaces $B_{p,q}^s(\mathbb{R}^d)$, $s \in \mathbb{R}$, $1 \leqslant p, q \leqslant \infty$, are (see [Triebel (1977), Umarov (2015b)])

$$B_{p,q}^s(\mathbb{R}^d) = \{f = \sum_{j=0}^{\infty} a_j(x) : a_j \in L^p(\mathbb{R}^d),\ \mathrm{supp} F[a_0] \subset \{|\xi| \leqslant 2\}, \tag{4.4}$$

$$\mathrm{supp} F[a_j] \subset \{2^{j-1} \leqslant |\xi| \leqslant 2^{j+1}\},\ j = 1, 2, \ldots,\ \text{and} \tag{4.5}$$

$$\|f\|_{B_{p,q}^s}^q = \sum_{j=0}^{\infty} 2^{sjq} \|a_j\|_{L^p}^q < \infty\}. \tag{4.6}$$

Pseudo-differential operators in the spaces of tempered distributions, Schwartz distributions and distributions with compact support, can be defined by duality, namely,

$$AF(f) = F(A^* f), \quad F \in \mathcal{G}',\ f \in \mathcal{G}, \tag{4.7}$$

$$AF(f) = F(A^* f), \quad F \in \mathcal{D}',\ f \in \mathcal{D}, \tag{4.8}$$

$$AF(f) = F(A^* f), \quad F \in \mathcal{E}',\ f \in \mathcal{E}, \tag{4.9}$$

where A^* is the adjoint to the pseudo-differential operator A, and $F(f)$ means the value of the distribution F on the element f.

Proposition 4.1.

 (1) *Let $A \in OPS^m_{\rho,\delta}(\Omega)$. Then the mapping $A : \mathcal{D}(\Omega) \to \mathcal{E}(\Omega)$ and, by duality, the mapping $A : \mathcal{E}'(\Omega) \to \mathcal{D}'(\Omega)$ are continuous.*

 (2) *Let $A \in OPS^m_{\rho,\delta}$, $\delta < 1$. Then the mapping $A : \mathcal{G}(\mathbb{R}^d) \to \mathcal{G}(\mathbb{R}^d)$ and, by duality, the mapping $A : \mathcal{G}'(\mathbb{R}^d) \to \mathcal{G}'(\mathbb{R}^d)$ are continuous.*

 (3) *Let $A \in OPS^m_{\rho,\delta}$, $0 \leqslant \delta < \rho \leqslant 1$. Then the mapping $A : H^s(\mathbb{R}^d) \to H^{s-m}(\mathbb{R}^d)$ is continuous for all $s \in \mathbb{R}$.*

For the mapping properties of pseudo-differential operators on the generalized Sobolev, Besov, and other useful spaces, see [Triebel (1977), Taylor (1981), Shubin (2001), Hörmander (2007), Umarov (2015b)].

Some fractional order differential operators studied in the previous chapter are pseudo-differential operators. Below we reformulate some results presented in the previous chapter in terms of pseudo-differential operators.

Proposition 4.2. *Let $\alpha > 0$. The operators $_{-\infty}D^\alpha$ and $_xD^\alpha_\infty$, $\mathbb{R}^\alpha_0$ and $\mathbb{D}^\alpha_0$, defined in Sections 3.5 and 3.7 respectively, are pseudo-differential operators with the following symbols:*

 (1) $\sigma_{_{-\infty}D^\alpha}(\xi) = (i\xi)^\alpha$, $\xi \in \mathbb{R}$;

 (2) $\sigma_{_xD^\alpha_\infty}(\xi) = (-i\xi)^\alpha$, $\xi \in \mathbb{R}$;

 (3) $\sigma_{\mathbb{R}^\alpha_0}(\xi) = |\xi|^{-\alpha}$, $\xi \in \mathbb{R}^d$;

 (4) $\sigma_{\mathbb{D}^\alpha_0}(\xi) = -|\xi|^\alpha$, $\xi \in \mathbb{R}^d$.

4.2   Pseudo-differential operators with singular symbols

Symbols of pseudo-differential operators associated with stochastic processes driven by time-changed Lévy processes, in general, do not satisfy condition (4.1), and hence, do not belong to Hörmander classes. Moreover, these symbols, in general, are not differentiable in the dual variable. Pseudo-differential operators with symbols singular in the dual variable are studied in the works [Dubinskii (1991), Umarov (2015b)]. Below we briefly expose some main properties of such operators.

It is not hard to verify that if the symbol has a nonintegrable singularity, then the corresponding operator may not be meaningful even on infinitely differentiable functions with compact support. Here is an example. Let the symbol $a(x, \xi)$, $(x, \xi) \in \Omega \times \mathbb{R}^d$, have a nonintegrable singularity at $\xi_0 \in \mathbb{R}^d$. Let $f(x) \in C^\infty_0(\Omega)$ have the Fourier transform $F[f](\xi)$ satisfying the condition $F[f](\xi) = 1$ in a neighborhood of ξ_0. It is easy to construct such a function. Now using definition (4.2) one can see that the pseudo-differential operator $A(x, D)$ corresponding to the symbol $a(x, \xi)$, is not well-defined on f, that is, $|A(x, D)f(x)| = \infty$. Therefore, it is important to have an appropriate domain for pseudo-differential operators with singular symbols. One possible space of distributions, called ψ-distributions, is the space defined below and denoted by $\Psi'_G(\mathbb{R}^n)$, $G \subset \mathbb{R}^d$. The topological and other properties of this and more general spaces of ψ-distributions are systematically provided in the book [Umarov (2015b)].

Let G be a subset of $\mathbb{R}^d$ and suppose that all the singularities of the symbol $a(x,\xi)$ with respect to ξ are concentrated on $\mathbb{R}^d\backslash G$. Denote by $\Psi_G(\mathbb{R}^d)$ the set of functions $f \in L^2(\mathbb{R}^d)$ such that the support of f is a compact subset of G, that is

$$\text{supp}\,[f] \subset G \quad \text{and} \quad \text{supp}\,[f] \text{ is compact.} \tag{4.10}$$

We say that a sequence $f_m \in \Psi_G(\mathbb{R}^d)$, $m = 1, 2, \ldots$, converges to $f \in \Psi_G(\mathbb{R}^d)$ if

(1) there exists a compact set $K \subset G$ such that $\text{supp}[f_m] \subset K$ for all $m = 1, 2, \ldots$, and
(2) $f_m \to f$ as $m \to \infty$ in the norm of $L^2(\mathbb{R}^d)$.

The space $\Psi_G(\mathbb{R}^d)$ with the topology defined with the help of this convergence is a locally convex topological vector space [Umarov (2015b)]. Denote by $\Psi'_G(\mathbb{R}^d)$ the space of distributions defined on $\Psi_G(\mathbb{R}^d)$ with the weak topology. Elements of this space are called ψ-distributions. One can represent $\Psi_G(\mathbb{R}^d)$ as an inductive limit of Banach spaces with the finest topology and $\Psi'_G(\mathbb{R}^d)$ as a projective limit of the dual Banach spaces with the coarsest topology; see details in [Umarov (2015b)]. The duality of theses spaces will be denoted by $\langle \cdot, \cdot \rangle$. Namely, if $F \in \Psi'_G(\mathbb{R}^d)$ and $f \in \Psi_G(\mathbb{R}^d)$, then the value of F at f is $F(f) = \langle F, f \rangle$.

Taking condition (4.10) into account one can define a pseudo-differential operator with the symbol $a(x,\xi)$ as follows:

$$A(x,D)f(x) = \frac{1}{(2\pi)^d} \int_G a(x,\xi)F[f](\xi)e^{-i(x,\xi)}d\xi, \quad x \in \Omega, \tag{4.11}$$

where $f \in \Psi_G(\mathbb{R}^d)$. The integral here is meaningful since singularities of the symbol $a(x,\xi)$ are outside of the set G. Pseudo-differential operators acting on ψ-distributions can be defined in the standard way (see (4.7)–(4.9)), namely,

$$\langle A(x,D)F, f \rangle = \langle F, A(x,-D)f \rangle, \quad F \in \Psi'_G(\mathbb{R}^d), \ f \in \Psi_{-G}(\mathbb{R}^d), \tag{4.12}$$

where $A(x,-D)$ is the pseudo-differential operator with the symbol $a(x,-\xi)$ and $-G = \{-\xi; \xi \in G\}$.

We have mentioned above that symbols of pseudo-differential operators arising in the description of some stochastic processes are not differentiable but are continuous in the dual variable ξ. Therefore, we will assume that the symbol $a(x,\xi)$ is continuous in the variable $\xi \in G$. Without loss of generality we can assume that dependence of the symbol on the variable x is smooth. Further, it is also assumed that the singularity set of the symbol is not too large. Namely, we assume that the d-dimensional Lebesgue measure of $\mathbb{R}^d\backslash G$ is zero. This requirement is crucial due to the following denseness result:

Proposition 4.3. [Umarov (2015b)] $\Psi_G(\mathbb{R}^d)$ *is densely embedded into* $L^2(\mathbb{R}^d)$ *if and only if* $\mathbb{R}^d\backslash G$ *has d-dimensional measure zero.*

Now we define a wide class of symbols without any growth conditions for large ξ and for which a meaningful theory of pseudo-differential operators can be constructed.

Definition 4.1. Let $a(x, \xi)$ be continuous in $\xi \in G$, infinitely differentiable in $x \in \Omega$, and it may have any type of singularity on the boundary of G or outside of G. We denote this class of symbols by $S^0(G)$. Similarly, one can define the classes of symbols $S^m(G)$ and $S^\infty(G)$ if $a(x, \xi)$ is m times differentiable and infinitely differentiable in $\xi \in G$, respectively.

Proposition 4.4. [Umarov (2015b)] *Let $a(x, \xi) \in S^0(G)$ and A be the pseudo-differential operator with the symbol $a(x, \xi)$.*

(1) *Then the mapping $A : \Psi_G(\mathbb{R}^d) \to \mathcal{E}(\mathbb{R}^d)$ is continuous;*

(2) *If $a(x, \xi)$ has a compact support in the variable x, then the mapping $A : \Psi_G(\mathbb{R}^d) \to \mathcal{D}(\mathbb{R}^d)$ is continuous;*

(3) *If the Fourier transform of $a(x, \xi)$ in the variable x has a compact support K for all $\xi \in G$, then the mapping $A : \Psi_G(\mathbb{R}^d) \to \Psi_{G+K}(\mathbb{R}^d)$ is continuous. Here $G + K = \{\xi + \eta; \xi \in G, \eta \in K\}$.*

By duality, the following proposition also holds.

Proposition 4.5. [Umarov (2015b)] *Let $a(x, \xi) \in S^0(G)$ and A be the pseudo-differential operator with the symbol $a(x, \xi)$.*

(1) *Then the mapping $A : \mathcal{E}'(\mathbb{R}^d) \to \Psi'_{-G}(\mathbb{R}^d)$ is continuous;*

(2) *If $a(x, \xi)$ has a compact support in the variable x, then the mapping $A : \mathcal{D}'(\mathbb{R}^d) \to \Psi'_{-G}(\mathbb{R}^d)$ is continuous;*

(3) *If the Fourier transform of $a(x, \xi)$ in the variable x has a compact support K for all $\xi \in G$, then the mapping $A : \Psi'_{-G-K}(\mathbb{R}^d) \to \Psi'_{-G}(\mathbb{R}^d)$ is continuous.*

A pseudo-differential operator with a symbol in $S^0(G)$ can also be extended to Sobolev spaces under certain conditions. Namely, the following proposition holds.

Proposition 4.6. [Umarov (2015b)] *Let $a(x, \xi) \in S^0(G)$ and assume the following conditions:*

(1) *the d-dimensional Lebesgue measure of the set $\mathbb{R}^d \backslash G$ is zero;*

(2) *the Fourier transform of $a(x, \xi)$ in the variable x has a compact support K for all $\xi \in G$;*

(3) *there exist a nonnegative function $k(u) \in L^1[0, \infty)$ and a number $m \in \mathbb{R}$ such that*

$$|a(x, \xi)| \leqslant k(|x|)(1 + |\xi|^2)^{m/2}, \quad x \in \Omega, \ \xi \in G.$$

Then for any $s \in \mathbb{R}$, the mapping $A : H^s(\mathbb{R}^d) \to H^{s-m}(\mathbb{R}^d)$ is continuous.

4.3   Pseudo-differential operators associated with Lévy processes

In this section we will deal with pseudo-differential operators directly connected with the Lévy processes (see Section 5.3). In general, these are second order pseudo-differential operators of the form

$$A\varphi(x) = c_0(x)\varphi(x) + \sum_{j=1}^{d} b_j(x)\frac{\partial\varphi(x)}{\partial x_j} + \sum_{j,k=1}^{d} a_{jk}(x)\frac{\partial^2\varphi(x)}{\partial x_j \partial x_k}$$

$$+ \int_{\mathbb{R}^d\setminus\{0\}} \left[\varphi(x+y) - \varphi(x) - \frac{(\nabla\varphi(x), y)}{1+|y|^2}\right]\nu(x, dy), \qquad (4.13)$$

where $c_0 : \mathbb{R}^d \to \mathbb{R}$, $b = (b_1(x), \ldots, b_d(x)) : \mathbb{R}^d \to \mathbb{R}^d$, $a = (a_{jk}(x))_{jk=1}^{d} : \mathbb{R}^d \to \mathbb{R}^{d\times d}$ are measurable mappings satisfying some continuity and growth conditions specified below, and $\nu(x, \cdot)$ is a Borel measure defined on $\mathbb{R}^d\setminus\{0\}$ and satisfying the condition

$$\int_{\mathbb{R}^d\setminus\{0\}} \min(1, |y|^2)\nu(x, dy) < \infty,$$

for all $x \in \mathbb{R}^d$. Such a measure is called a *Lévy measure*. We will see in Section 5.3 that the first three terms in (4.13) are associated with the Brownian component of the process, and the integral term is associated with jump components. If the mappings c_0, b and a are zero-mappings, then the corresponding Lévy process is a pure jump process. Note that representation (4.13) of the pseudo-differential operator A is equivalent to

$$A\varphi(x) = c_0(x)\varphi(x) + \sum_{j=1}^{d} b_j(x)\frac{\partial\varphi(x)}{\partial x_j} + \sum_{j,k=1}^{d} a_{jk}(x)\frac{\partial^2\varphi(x)}{\partial x_j \partial x_k}$$

$$+ \int_{\mathbb{R}^d\setminus\{0\}} \left[\varphi(x+y) - \varphi(x) - (\nabla\varphi(x), y)\, \boldsymbol{I}_{|y|\leqslant 1}\right]\nu(x, dy), \qquad (4.14)$$

where $\boldsymbol{I}_{|y|\leqslant 1}$ is the indicator function of the unit ball. In general, the functions $(1 + |y|^2)^{-1}$ in representation (4.13) and $\boldsymbol{I}_{|y|\leqslant 1}$ in representation (4.14) can be replaced by a smooth function $\phi(y)$ with a compact support and identically equal to 1 in a neighborhood of zero. Later we introduce a function $\phi(x, y)$ having such properties in a neighborhood of each $x \in \mathbb{R}^d$, which will be called a local unit function.

Operators of the form (4.13) (or (4.14)) are called *Lévy-type operators*. Some authors call only the integral part

$$S\varphi(x) = \int_{\mathbb{R}^d\setminus\{0\}} \left[\varphi(x+y) - \varphi(x) - (\nabla\varphi(x), y)\, \boldsymbol{I}_{|y|\leqslant 1}\right]\nu(x, dy)$$

of the operator A a *Lévy operator,* calling the differential part

$$R\varphi(x) = (A - S)\varphi(x) = c_0(x)\varphi(x) + \sum_{j=1}^{d} b_j(x)\frac{\partial\varphi(x)}{\partial x_j} + \sum_{j,k=1}^{d} a_{jk}\frac{\partial^2\varphi(x)}{\partial x_j \partial x_k}$$

a *diffusion operator.* Consider some examples of Lévy-type operators.

Example 4.1.

1. *Operators associated with Brownian motion.* Let ν be the zero-measure and $c_0(x) \equiv 0$. Then the operator A in (4.13) or in (4.14) represents the Fokker-Planck operator

$$A\varphi(x) = \sum_{j=1}^{d} b_j(x) \frac{\partial \varphi(x)}{\partial x_j} + \sum_{j,k=1}^{d} a_{jk}(x) \frac{\partial^2 \varphi(x)}{\partial x_j \partial x_k}$$

   associated with Brownian motion with covariance matrix $\mathbb{A}(x) = (a_{jk}(x))$ and drift $b(x) = (b_1(x), \ldots, b_d(x))$.

2. *Operators associated with pure jump processes.* Let $c_0(x) \equiv 0$, $a_{jk}(x) \equiv 0$, $b_j(x) \equiv 0$, and $\nu(x, dy) = \nu(dy)$, a Lévy measure not depending on x. Then the operator A in (4.13) takes the form

$$A\varphi(x) = \int_{\mathbb{R}^d \setminus \{0\}} \left[\varphi(x+y) - \varphi(x) - \frac{(\nabla \varphi(x), y)}{1 + |y|^2} \right] \nu(dy). \tag{4.15}$$

   This is an operator associated with a pure jump process determined by the Lévy measure ν. In fact, the operator A in (4.15) is a pseudo-differential operator whose symbol is

$$\sigma_A(\xi) = \int_{\mathbb{R}^d \setminus \{0\}} \left[e^{i(\xi, y)} - 1 - \frac{i(\xi, y)}{1 + |y|^2} \right] \nu(dy). \tag{4.16}$$

   Indeed, applying the Fourier transform to the integral in equation (4.15) yields

$$F[A\varphi](\xi) = \int_{\mathbb{R}^d \setminus \{0\}} \left[e^{i(\xi, y)} - 1 - \frac{i(\xi, y)}{1 + |y|^2} \right] F[\varphi](\xi) \nu(dy).$$

   Now applying the inverse Fourier transform to the latter yields

$$A\varphi(x) = \frac{1}{(2\pi)^d} \int_{\mathbb{R}^d} \int_{\mathbb{R}^d \setminus \{0\}} \left[e^{i(\xi, y)} - 1 - \frac{i(\xi, y)}{1 + |y|^2} \right] \nu(dy) F[\varphi](\xi) e^{-i(x, \xi)} d\xi,$$

   confirming that A is a pseudo-differential operator with the symbol $\sigma_A(\xi)$ defined in (4.16).

3. *Operators associated with spherically symmetric α-stable Lévy processes.* Let $c_0(x) \equiv 0$, $a_{jk}(x) \equiv 0$, $b_j(x) \equiv 0$, and $\nu(dy) = C_\alpha |y|^{-d-\alpha} dy$, where $0 < \alpha < 2$ and C_α is a constant specified below. Then the operator A in (4.14) takes the form

$$A\varphi(x) = C_\alpha \int_{\mathbb{R}^d \setminus \{0\}} \left[\varphi(x+y) - \varphi(x) - (\nabla \varphi(x), y) \, \boldsymbol{I}_{|y| \leq 1} \right] \frac{dy}{|y|^{d+\alpha}}. \tag{4.17}$$

One can show that in this case the symbol of the operator A with the appropriate constant C_α is represented as $\sigma_A(\xi) = -|\xi|^\alpha$. The corresponding operator A is associated with the so-called spherically symmetric α-stable Lévy process (see Section 5.3). To show that $\sigma_A(\xi) = -|\xi|^\alpha$, first notice that as in (4.16), the symbol of A has the form

$$\sigma_A(\xi) = C_\alpha \int_{\mathbb{R}^d \setminus \{0\}} \left[e^{i(\xi, y)} - 1 - i(\xi, y) \, \boldsymbol{I}_{|y| \leq 1} \right] \frac{dy}{|y|^{d+\alpha}}. \tag{4.18}$$

We rewrite equation (4.18) in the form

$$\sigma_A(\xi) = C_\alpha \int_{|y| \leqslant 1} \frac{e^{i(y,\xi)} - 1 - i(y,\xi)}{|y|^{\alpha+d}} dy + C_\alpha \int_{|y|>1} \frac{e^{i(y,\xi)} - 1}{|y|^{\alpha+d}} dy.$$

Using the substitution $y = x/|\xi|$, $\xi \neq 0$, in the integrals yields

$$\sigma_A(\xi) = C_\alpha |\xi|^\alpha \left(\int_{|x| \leqslant |\xi|} \frac{e^{i(x,\theta)} - 1 - i(x,\theta)}{|x|^{\alpha+d}} dx + \int_{|x|>|\xi|} \frac{e^{i(x,\theta)} - 1}{|x|^{\alpha+d}} dx \right),$$

where θ is a point on the unit sphere in $\mathbb{R}^d$ with the center at the origin, and therefore, the expression in parentheses does not depend on θ. Further, taking into account the equality

$$\int_{\mathbb{R}^d} \frac{(x,\theta)(\boldsymbol{I}_{|x| \leqslant 1} - \boldsymbol{I}_{|x| \leqslant |\xi|})}{|x|^{d+\alpha}} dx = 0,$$

one obtains

$$\sigma_A(\xi) = C_\alpha |\xi|^\alpha \int_{\mathbb{R}^d} \left(e^{i(x,\theta)} - 1 - i(x,\theta)\boldsymbol{I}_{|x| \leqslant 1} \right) \frac{dx}{|x|^{\alpha+d}} = -|\xi|^\alpha,$$

where we set

$$C_\alpha = - \left[\int_{\mathbb{R}^d} \left(e^{i(x,\theta)} - 1 - i(x,\theta)\boldsymbol{I}_{|x| \leqslant 1} \right) \frac{dx}{|x|^{\alpha+d}} \right]^{-1}.$$

4. *Operators associated with general α-stable Lévy processes.* Pseudo-differential operators associated with α-stable Lévy processes have symbols

$$\sigma(\xi) = i(b,\xi) - \int_{S^{d-1}} |(\xi,\theta)|^\alpha \left[1 - i \tan\left(\frac{\pi\alpha}{2}\right) \mathrm{sign}(\xi,s) \right] \rho(ds)$$

if $0 < \alpha < 2$, $\alpha \neq 1$, and

$$\sigma(\xi) = i(b,\xi) - \int_{S^{d-1}} |(\xi,\theta)| \left[1 + i\frac{2}{\pi} \ln |(\xi,s)| \mathrm{sign}(\xi,s) \right] \rho(ds),$$

if $\alpha = 1$, where S^{d-1} is the $(d-1)$-dimensional unit sphere, and ρ is a finite measure on S^{d-1}.

4.4 Some abstract facts on semigroups and linear operators

Below we provide (without proof) some known and important properties of the pseudo-differential operators associated with Lévy and Lévy-type processes. We refer the reader for details to the following sources: [Applebaum (2009), Barndorf-Nielsen et al. (2001), Jacob (2001, 2002, 2005), Hoh (2000), Situ (2005)]. We begin with some definitions.

Definition 4.2. Let $\mathcal{X}$ be a Banach space and $\{T_t, \ t \geqslant 0\}$ be a one-parameter family of linear operators mapping $\mathcal{X}$ to itself. The family $\{T_t\}$ is called a *strongly continuous semigroup* if

(1) $T_0 = I$, the identity operator,

(2) $T_t T_s = T_{t+s}$ for all $t, s \geqslant 0$, and

(3) for any $t_0 \geqslant 0$, the convergence $T_t \varphi \to T_{t_0} \varphi$ in the norm of $\mathcal{X}$ holds as $t \to t_0$ for all $\varphi \in \mathcal{X}$.

Definition 4.3. A linear operator A defined as

$$A\varphi = \lim_{t \to 0+} \frac{T_t \varphi - \varphi}{t}, \tag{4.19}$$

provided the limit in the norm of $\mathcal{X}$ exists, is called the *infinitesimal generator* of the semigroup $\{T_t\}$. In fact, the set of elements $\varphi \in \mathcal{X}$ for which the limit (4.19) exists is a dense subset of $\mathcal{X}$ and is the domain of the operator A. We will denote the domain of A by $\mathrm{Dom}(A)$. A subset of $\mathrm{Dom}(A)$ dense in $\mathcal{X}$ is called a *core* of the infinitesimal generator A.

Proposition 4.7. [Engel and Nagel (1999)] *For any strongly continuous semigroup* $\{T_t, \ t \geqslant 0\}$, *there exist numbers* $\omega \in \mathbb{R}$ *and* $M \geqslant 1$ *such that*

$$\|T_t \varphi\| \leqslant M e^{\omega t} \|\varphi\| \tag{4.20}$$

for all $t \geqslant 0$ *and* $\varphi \in \mathcal{X}$.

Let $\varphi \in \mathcal{X}$. Consider an element $u(t) = u_\varphi(t) = T_t \varphi$. The mapping $u_\varphi(\cdot) : \mathbb{R}_+ \to \mathcal{X}$ is called an *orbit map*. It follows immediately from Proposition 4.20 that any orbit map $u_\varphi(\cdot)$ is locally bounded and continuous at every point $t \geqslant 0$.

Definition 4.4. The strongly continuous semigroup is called

(1) *bounded* if $\omega = 0$ in (4.20);

(2) *contractive* if $\omega = 0$ and $M = 1$ in (4.20);

(3) *isometric* if $\|T_t \varphi\| = \|\varphi\|$ for all $t \geqslant 0$ and $\varphi \in \mathcal{X}$.

Proposition 4.8. [Engel and Nagel (1999)] *Let an operator* A *with a domain* $\mathrm{Dom}(A)$ *be the infinitesimal generator of a strongly continuous semigroup* $\{T_t, \ t \geqslant 0\}$. *Then*

(1) A *is a linear closed operator;*

(2) *for each* $\varphi \in \mathcal{X}$ *the vector function* $u(t) = T_t \varphi$ *is differentiable in the norm of* $\mathcal{X}$;

(3) *if* $\varphi \in \mathrm{Dom}(A)$, *then* $u(t) = T_t \varphi \in \mathrm{Dom}(A)$ *and*

$$\frac{du(t)}{dt} = Au(t).$$

Proposition 4.9. [Engel and Nagel (1999)] *Let* $\{T_t, \ t \geqslant 0\}$ *be a strongly continuous semigroup satisfying estimate (4.20) and with the infinitesimal generator* A *with the domain* $\mathrm{Dom}(A)$. *Then*

(1) *a complex number* $\lambda \in \mathbb{C}$ *such that* $\mathrm{Re}(\lambda) > \omega$ *belongs to the resolvent set of* A *and the resolvent operator* $R_A(\lambda) = (A - \lambda I)^{-1}$ *has the representation*

$$R_A(\lambda)(\varphi) = \int_0^\infty e^{-\lambda s} T_s \varphi \, ds,$$

for all $\varphi \in \mathcal{X}$;

(2) *for the resolvent operator the following estimate holds:*

$$\|R_A(\lambda)\| \leqslant \frac{M}{Re(\lambda) - \omega}.$$

Example 4.2. Let $\mathcal{X} = L^2(\mathbb{R}^d)$ and

$$A = \frac{1}{2}\Delta = \frac{1}{2}\left(\frac{\partial^2}{\partial x_1^2} + \cdots + \frac{\partial^2}{\partial x_d^2}\right)$$

with the domain $\mathrm{Dom}(A) = H^1(\mathbb{R}^d)$, the Sobolev space defined in Section 4.1. This operator is the infinitesimal generator of the strongly continuous semigroup $\{T_t, \ t \geqslant 0\}$ defined by

$$T_t\varphi(x) = \frac{1}{(2\pi t)^{d/2}} \int_{\mathbb{R}^d} e^{-\frac{|x-y|^2}{2t}} \varphi(y)dy, \quad t \geqslant 0, \ x \in \mathbb{R}^d. \tag{4.21}$$

In fact, the right hand side of (4.21) is the convolution $T_t\varphi(x) = (G_t * \varphi)(x)$, where

$$G_t(x) = \frac{1}{(2\pi t)^{d/2}} e^{-\frac{|x|^2}{2t}}, \quad t \geqslant 0, \ x \in \mathbb{R}^d, \tag{4.22}$$

is the d-dimensional Gaussian density evolved in time. Here the symbol "$*$" stands for the convolution operation, which is defined for functions $f, g \in L^2(\mathbb{R}^d)$ by

$$(f * g)(x) = \int_{\mathbb{R}^d} f(y)g(x-y)dy, \quad x \in \mathbb{R}^d.$$

Semigroups defined through the convolution are called *convolution semigroups*. We note that the semigroup $\{T_t, \ t \geqslant 0\}$ in equation (4.21) can also be represented as the expectation

$$T_t\varphi(x) = \mathbb{E}[\varphi(x + B_t)]$$
$$= \frac{1}{(2\pi t)^{d/2}} \int_{\mathbb{R}^d} \varphi(x + y)e^{-\frac{|y|^2}{2t}} dy, \quad t \geqslant 0, \ x \in \mathbb{R}^d,$$

where $(B_t)_{t \geqslant 0}$ is d-dimensional Brownian motion starting at 0. This immediately follows from the fact that the density function of B_t for each fixed t is the Gaussian density given in (4.22).

This example shows a connection between Brownian motion and a strongly continuous convolution semigroup. Indeed, there is a deep connection between Markovian stochastic processes and strongly continuous semigroups. Below we will describe a wide class of pseudo-differential operators, which are infinitesimal generators of strongly continuous semigroups linked with Markovian stochastic processes. In the following, we use the notations $C_b(\mathbb{R}^d)$, $C_0(\mathbb{R}^d)$, and $C_c(\mathbb{R}^d)$ to respectively denote the spaces of bounded continuous functions on $\mathbb{R}^d$, continuous functions on $\mathbb{R}^d$ vanishing at infinity, and continuous functions on $\mathbb{R}^d$ with compact support. The spaces $C_0^\infty(\mathbb{R}^d)$ and $C_c^\infty(\mathbb{R}^d)$ are defined in a similar manner using infinitely differentiable functions.

Proposition 4.10. *Let* $(L_t)_{t \geqslant 0}$ *be a Lévy process with triple* (b, Σ, ν) *(to be defined in Section 5.3). Then*

(1) $T_t\varphi(x) = \mathbb{E}[\varphi(x + L_t)]$ *is a strongly continuous semigroup in the space* $\mathcal{X} = C_b(\mathbb{R}^d)$ *with the* sup-*norm;*

(2) *the corresponding infinitesimal generator* A *is defined by the pseudo-differential operator*

$$Af(x) = \sum_{j=1}^{d} b_j \frac{\partial f}{\partial x_j} + \sum_{i,j=1}^{d} \sigma_{i,j} \frac{\partial^2 f}{\partial x_i \partial x_j}$$

$$+ \int_{\mathbb{R}^d \setminus \{0\}} \left[f(x + w) - f(x) - I_{|w| \leqslant 1} \sum_{j=1}^{d} w_j \frac{\partial f(x)}{\partial x_j} \right] \nu(dw), \qquad (4.23)$$

with the core $C_0^\infty(\mathbb{R}^d)$.

Definition 4.5. Let A be a linear closed operator in the space $\mathcal{X} = C_0(\mathbb{R}^d)$ with domain $\mathrm{Dom}(A)$. The operator A is said to satisfy the positive maximum principle if $\varphi \in \mathrm{Dom}(A)$ and $\varphi(x_0) = \sup_{x \in \mathbb{R}^d} \varphi(x) \geqslant 0$ implies $(A\varphi)(x_0) \leqslant 0$.

Definition 4.6. A semigroup $\{T_t,\ t \geqslant 0\}$ defined in $\mathcal{X} = C_0(\mathbb{R}^d)$ is called a *positivity preserving semigroup* if $T_t\varphi(x) \geqslant 0$ for all $t \geqslant 0$ and all $x \in \mathbb{R}^d$ whenever $\varphi(x) \geqslant 0$ for all $x \in \mathbb{R}^d$.

Definition 4.7. A strongly continuous, positivity preserving, contraction semigroup $T_t : C_0(\mathbb{R}^d) \to C_0(\mathbb{R}^d)$ is called a *Feller semigroup.*

The following theorem characterizes infinitesimal generators of Feller semigroups.

Theorem 4.1 (Hille–Yosida–Ray). *A closed operator* A *with a dense domain is the infinitesimal generator of a strongly continuous, positivity preserving, contraction semigroup on* $C_0(\mathbb{R}^d)$ *if and only if the following conditions are verified:*

(1) $(0, \infty) \subset \rho(A)$, *where* $\rho(A)$ *is the resolvent set of* A;

(2) A *satisfies the positive maximum principle.*

Definition 4.8. [Applebaum (2009)] A *Lévy kernel* is a family $\{\nu(x, \cdot),\ x \in \mathbb{R}^d\}$, where each $\nu(x, \cdot)$ is a Borel measure on $\mathbb{R}^d \setminus \{x\}$, such that

(1) the mapping $x \to \int_{\mathbb{R}^d \setminus \{x\}} |y - x|^2 f(y) \nu(x, dy)$ is Borel measurable and locally bounded for each $f \in C_c(\mathbb{R}^d)$;

(2) for each $x \in \mathbb{R}^d$, and for every neighborhood U_x of x, the boundedness $\nu(x, \mathbb{R}^d \setminus U_x) < \infty$ holds.

Definition 4.9. A *local unit function* is an $L^1(\mathbb{R}^d \times \mathbb{R}^d)$-function $\phi(x, y) : \mathbb{R}^d \times \mathbb{R}^d \to [0, 1]$ such that $\phi(x, y) \equiv 1$ in a neighborhood of the diagonal $\{(x, y) \in \mathbb{R}^d \times \mathbb{R}^d : x = y\}$, and for every compact set $K \subset \mathbb{R}^d$, the functions $y \to \phi(x, y)$, $x \in K$, have their support in a fixed compact set.

Theorem 4.2 (Courrége's First Theorem). *If A is a linear operator in $C_0(\mathbb{R}^d)$ and $C_c^\infty(\mathbb{R}^d) \subset Dom(A)$, then A satisfies the positive maximum principle if and only if there exist*

(1) *continuous functions $c_0 : \mathbb{R}^d \to \mathbb{R}$ and $b_j : \mathbb{R}^d \to \mathbb{R}$, $j = 1, \ldots, d$, such that $c_0(x) \leqslant 0$ for all $x \in \mathbb{R}^d$,*

(2) *mappings $a_{ij} : \mathbb{R}^d \to \mathbb{R}$, $i, j = 1, \ldots, d$, such that $a(x) \equiv (a_{ij}(x))_{i,j=1,\ldots,d}$ is a positive symmetric matrix for each $x \in \mathbb{R}^d$ and the mapping $x \to (y, a(x)y)$ is upper semicontinuous for each $y \in \mathbb{R}^d$,*

(3) *a Lévy kernel ν, and*

(4) *a local unit function ϕ,*

such that for all $f \in C_c^\infty(\mathbb{R}^d)$ and $x \in \mathbb{R}^d$,

$$Af(x) = c_0(x)f(x) + \sum_{i=1}^d b_i(x)\frac{\partial f(x)}{\partial x_i} + \sum_{ij=1}^d a_{ij}(x)\frac{\partial^2 f(x)}{\partial x_i \partial x_j}$$

$$+ \int_{\mathbb{R}^d \setminus \{x\}} \left[f(y) - \phi(x,y)\left(f(x) + \sum_{i=1}^d (y_i - x_i)\frac{\partial f(x)}{\partial y_i} \right) \right] \nu(x, dy). \qquad (4.24)$$

Theorem 4.3 (Courrége's Second Theorem). *Let A be a linear operator in $C_0(\mathbb{R}^d)$. Suppose that $C_c^\infty(\mathbb{R}^d) \subset Dom(A)$ and that A satisfies the positive maximum principle. Define the mapping $a(x, \xi) = e^{i(x,\xi)} A e^{-i(x,\xi)}$. Then*

(1) *for each fixed $x \in \mathbb{R}^d$, $a(x, \xi)$ is continuous, hermitian and conditionally positive definite in the variable ξ;*

(2) *the estimate $|a(x, \xi)| \leqslant c(x)|\xi|^2$ is valid for all $x, \xi \in \mathbb{R}^d$ with locally bounded function $c(x)$;*

(3) *A is a pseudo-differential operator well defined for all $f \in C_c^\infty(\mathbb{R}^d)$ and $a(x, \xi)$ is its symbol.*

Conversely, if $a(x, \xi)$ is continuous, hermitian and conditionally positively definite in the variable ξ, then the pseudo-differential operator A with the symbol $a(x, \xi)$ satisfies the positive maximum principle.

Remark 4.1. (a) Notice that the form of the pseudo-differential operator given in (4.24) in Courrége's Second Theorem differs from the form (4.14) in Section 4.3. However, as the calculation below shows, these two forms, in fact, are equivalent.

Indeed, we have

$$Af(x) = c_0(x)f(x) + \sum_{i=1}^{d} b_i(x)\frac{\partial f(x)}{\partial x_i} + \sum_{ij=1}^{d} a_{ij}(x)\frac{\partial^2 f(x)}{\partial x_i \partial x_j}$$

$$+ \int_{\mathbb{R}^d \setminus \{x\}} \left[f(y) - \left(\phi(x,y) - 1 + 1\right)f(x) - \phi(x,y)\sum_{i=1}^{d}(y_i - x_i)\frac{\partial f(x)}{\partial y_i}\right]\nu(x, dy)$$

$$= c_1(x)f(x) + \sum_{i=1}^{d} b_i(x)\frac{\partial f(x)}{\partial x_i} + \sum_{ij=1}^{d} a_{ij}(x)\frac{\partial^2 f(x)}{\partial x_i \partial x_j}$$

$$+ \int_{\mathbb{R}^d \setminus \{x\}} \left[f(y) - f(x) - \phi(x,y)\sum_{i=1}^{d}(y_i - x_i)\frac{\partial f(x)}{\partial y_i}\right]\nu(x, dy),$$

where

$$c_1(x) = c_0(x) - \int_{\mathbb{R}^d \setminus \{x\}} (\phi(x,y) - 1)\nu(x, dy).$$

Now the substitution $z = y - x$ in the integral term of the latter reduces the operator A to the form in equation (4.14).

(b) Courrége's two theorems describe important properties of pseudo-differential operators associated with Lévy and Lévy-type processes. Proposition 4.10 shows that pseudo-differential operators associated with Lévy processes have the form (4.23). These pseudo-differential operators are particular cases of pseudo-differential operators presented in (4.24) and correspond to the case of constant coefficients b_j, $a_{i,j}, i, j = 1, \dots, d$, and the Lévy measure ν independent of the variable x. Therefore, pseudo-differential operators of the form (4.24) with the core $C_0^\infty(\mathbb{R}^d)$ represent more general Lévy-type stochastic processes and are called Lévy-type operators. Lévy-type stochastic processes can be interpreted as Lévy processes whose parameters (b, Σ, ν) change from point to point, that is, these parameters depend on the spatial variable x. Lévy-type operators, as follows from the Courrége's First Theorem, satisfy the positive maximum principle. Their symbols, as follows from the Courrége's Second Theorem, are continuous, hermitian, and conditionally positive definite. Further properties of such pseudo-differential operators can be found in [Courrége (1964), Jacob and Leopold (1993), Sato (1999), Hoh (2000), Jacob (2001, 2002, 2005), Applebaum (2009)]. Finally, we note that the pseudo-differential operator A defined in equation (4.24) and associated with a Lévy-type process in a bounded domain considered in Section 4.6 is called *a second order Waldenfels operator* (see, e.g. [Taira (1992)]).

4.5   Pseudo-differential operators on manifolds

In the context of this book, pseudo-differential operators on manifolds occur in the study of stochastic processes in a bounded domain with smooth boundary (to be discussed in Sections 4.6 and 7.7). In fact, such a domain can be considered as a subset of a smooth manifold of the same dimension without boundary. Therefore, in this section we briefly discuss the definition and basic properties of pseudo-differential operators on manifolds. For details, consult [Taylor (1981), Schulze (1991), Shubin (2001), Hörmander (2007)]. The readers whose main interests are in stochastic processes in the whole space $\mathbb{R}^d$ can skip this section on a first reading.

Let M be a d-dimensional paracompact manifold. This means that every open cover of M has a locally finite open refinement. In general, to define a pseudo-differential operator on a manifold, one can transfer open sets of the manifold to open sets of $\mathbb{R}^d$. In Section 4.1, we discussed pseudo-differential operators acting on functions defined on open subsets of $\mathbb{R}^d$. Here, using the inverse transform, we define desired pseudo-differential operators on a manifold. If local coordinates are determined by diffeomorphisms, then the invariance principle (see [Taylor (1981), Egorov, et. al. (2013)]) holds, leading to a correct definition of pseudo-differential operators on manifolds.

Proposition 4.11. *Let Ω and Ω' be two open sets of $\mathbb{R}^d$ and $\kappa : \Omega \to \Omega'$ be a diffeomorphism. Let κ^* be the mapping $C_0^\infty(\Omega') \to C_0^\infty(\Omega)$ defined at functions $f \in C_0^\infty(\Omega')$ by $\kappa^* f(x) = f(\kappa(x))$. Further, let $A(x, D) \in OPS^m(\Omega)$. Then the pseudo-differential operator $A'(y, D)$ defined by the commutative diagram*

$$
\begin{array}{ccc}
C_0^\infty(\Omega) & \xrightarrow{\ A(x,D)\ } & C_0^\infty(\Omega) \\
\kappa^* \uparrow & & \downarrow (\kappa^*)^{-1} \\
C_0^\infty(\Omega') & \xrightarrow{\ A'(y,D)\ } & C_0^\infty(\Omega'),
\end{array}
\tag{4.25}
$$

is an operator in the class $OPS^m(\Omega')$. Moreover, if $a(x, \xi)$ is the symbol of $A(x, D)$, then the symbol of $A'(y, D)$ satisfies

$$a'(y, \xi) = a(\kappa^{-1}(y), J(y)\xi) \qquad \mathrm{mod}\ (S^{m-1}(\Omega')), \tag{4.26}$$

where $J(y)$ is the transpose of the Jacobian of the mapping κ^{-1} at the point $y \in \Omega'$.

The essence of this proposition is that pseudo-differential operators defined on a manifold do not depend on a choice of local coordinates. Therefore, the definition of a pseudo-differential operator $A'(y, D)$ given by (4.25) as well as the pseudo-differential operator $A'(x, D)$ defined on a manifold (see (4.28) below) are meaningful. We note that for the principal symbol $a'_m(x, \xi)$ of the operator $A'(y, D)$ relation (4.26) takes the form

$$a'_m(y, \xi) = a_m(\kappa^{-1}(y), J(y)\xi). \tag{4.27}$$

Let M_x be an open neighborhood of $x \in M$. Consider a diffeomorphism

$$\eta : M_x \to U_x,$$

where U_x is a subset of the d-dimensional Euclidean space T_x^* tangent to M at x. This diffeomorphism establishes local coordinates $y \in \mathbb{R}^d$ with

$$y_j = \eta_j(x_1, \ldots, x_d), \ j = 1, \ldots, d,$$

of the manifold M in the neighborhood of each point $x \in M$. Further, let $\{U_\alpha, \phi_\alpha\}$ be a chart of the manifold M, that is, $\{U_\alpha\}$ is a locally finite open covering of M and $\phi_\alpha : U_\alpha \to \mathcal{O}_\alpha$ is a diffeomorphic mapping of $U_\alpha \subset M$ to $\mathcal{O}_\alpha \subset \mathbb{R}^d$. Denote by ϕ_α^* a mapping which transfers functions or distributions f defined on U_α to functions or distributions defined on $\mathcal{O}_\alpha$ as $\phi_\alpha^* f(x) = (f \circ \phi_\alpha)(x)$. Thus, formally a pseudo-differential operator $A'(x, D)$ on the manifold M with the symbol $a(x, \xi) \in S^m(M)$ can be defined as an operator defined locally for each $x \in M$ through the following commutative diagram:

$$
\begin{array}{ccc}
\mathcal{F}(\mathcal{O}_\alpha) & \xrightarrow{A(y,D)} & \mathcal{F}(\mathcal{O}_\alpha) \\
\phi_\alpha^* \uparrow & & \downarrow (\phi_\alpha^*)^{-1} \\
\mathcal{F}(U_\alpha) & \xrightarrow{A'(x,D)} & \mathcal{F}(U_\alpha),
\end{array}
\tag{4.28}
$$

where $A(y, D)$, $y \in \mathcal{O}_\alpha$ is a pseudo-differential operator with a symbol in $S^m(\mathcal{O}_\alpha)$ defined on $\mathcal{O}_\alpha \subset \mathbb{R}_y^d$; see Section 4.1 for the definition of $S^m(\mathcal{O}_\alpha)$. The spaces $\mathcal{F}(\mathcal{O}_\alpha)$ are appropriate spaces of functions (or distributions) defined on the manifold M (discussed below). This definition of pseudo-differential operators on manifolds is meaningful since it does not depend on the choice of local coordinates due to Proposition 4.11. We use the notation $OPS^m(M)$ for pseudo-differential operators defined on a manifold M.

Appropriate spaces $\mathcal{F}(\cdot)$ in the definition (4.28) include the space of infinitely differentiable functions $C^\infty(M)$ or Sobolev spaces $H^s(M)$ of functions defined on the manifold M, or corresponding distributions. Let (U_α, ϕ_α) be a chart of the manifold M. Then the space $C^\infty(M)$ consists of functions f such that $\psi_\alpha \circ f \circ \phi_\alpha^{-1} \in C^\infty(\mathbb{R}^d)$. Here $\{\psi_\alpha(x)\}_{\alpha=1}^\infty$ is a smooth partition of unity corresponding to the covering $\{U_\alpha\}_{\alpha=1}^\infty$. In a similar manner, one can define the space $C_c^\infty(M)$ of functions vanishing outside of a compact set, Sobolev spaces $H^s(M)$, spaces of Bessel potentials $H_p^s(M)$, and Besov spaces $B_{p,q}^s(M)$ as the collections of functions f such that respectively $\psi_\alpha \circ f \circ \phi_\alpha^{-1} \in C_c^\infty(\mathbb{R}^d)$, $\psi_\alpha \circ f \circ \phi_\alpha^{-1} \in H^s(\mathbb{R}^d)$, $\psi_\alpha \circ f \circ \phi_\alpha^{-1} \in H_p^s(\mathbb{R}^d)$, and $\psi_\alpha \circ f \circ \phi_\alpha^{-1} \in B_{p,q}^s(\mathbb{R}^d)$. Topologies or norms of these spaces are induced from the corresponding topologies or norms of the corresponding spaces in local coordinates. We note that the Sobolev space $H^s(M)$ has a Hilbert structure induced from the Hilbert structure of $H_2^s(\mathbb{R}^n)$.

Below we provide mapping properties of pseudo-differential operators in the class $OPS^m(M)$.

Theorem 4.4. *Let a pseudo-differential operator $A(x, D) \in OPS^m(M)$ and M be a smooth paracompact manifold. Then the following mappings are continuous:*

(1) $A(x, D) : C_0^\infty(M) \to C^\infty(M)$;

(2) $A(x, D) : \mathcal{E}'(M) \to \mathcal{D}'(M)$;

(3) $A(x, D) : H_p^s(M) \to H_p^{s-m}(M)$, $s \in \mathbb{R}$, $1 \leqslant p \leqslant \infty$;

(4) $A(x, D) : B_{p,q}^s(M) \to B_{p,q}^{s-m}(M)$, $s \in \mathbb{R}, 1 \leqslant p, q \leqslant \infty$.

4.6 Pseudo-differential operators associated with stochastic processes in bounded domains

In Section 2.4.5, FPK equations associated with an SDE in a bounded domain $\Omega \subset \mathbb{R}^d$ with absorbing or reflecting boundary $\partial\Omega$ were considered. Facts provided in this section will be used in Section 7.7. Recall that the (backward) FPK equation associated with a homogeneous SDE is given by the initial-boundary value problem

$$\frac{\partial u(t, x)}{\partial t} = \sum_{j=1}^d b_j(x) \frac{\partial u(t, x)}{\partial x_j} + \sum_{i,j=1}^d a_{i,j}(x) \frac{\partial^2 u(t, x)}{\partial x_i \partial x_j} \quad t > 0, \ x \in \Omega, \tag{4.29}$$

$$\mathcal{B}u(t, x') \equiv \mu(x') \frac{\partial u(t, x')}{\partial \mathbf{n}} + \nu(x') u(t, x') = 0, \quad t > 0, \ x' \in \partial\Omega, \tag{4.30}$$

$$u(0, x) = \delta_0(x), \quad x \in \Omega, \tag{4.31}$$

where $b_j(x)$, $j = 1, \ldots, d$, are drift coefficients, $a_{i,j}(x)$, $i, j = 1, \ldots, d$, are diffusion coefficients, and the functions $\mu(x')$ and $\nu(x')$ are continuous functions defined on the boundary $\partial\Omega$. It is well-known that the stochastic process solving the SDE corresponding to the FPK equation in (4.29)–(4.31) is continuous. Such a stochastic process serves as a model of the diffusion process of a Markovian particle.

The FPK equation (4.29)–(4.31) does not take into account events with jumps and some specific phenomena which may occur on the boundary (diffusion on the boundary, jumps on the boundary or into the domain, viscosity, etc.). The general case with jumps and viscosity effects involves pseudo-differential operators in the FPK equation and in the boundary condition. As is noted above, the symbols of pseudo-differential operators associated with Lévy processes, including those in bounded domains which are considered in this section, are continuous but not necessarily smooth.

Let $\Omega \subset \mathbb{R}^d$ be a bounded domain with a smooth boundary $\partial\Omega \subset \mathbb{R}^{d-1}$. In the general case, the (backward) FPK equation can be formulated as follows:

$$\frac{\partial u(t, x)}{\partial t} = A(x, D)u(t, x), \quad t > 0, \ x \in \Omega, \tag{4.32}$$

$$\mathcal{W}(x', D)u(t, x') = 0, \quad t > 0, \ x' \in \partial\Omega, \tag{4.33}$$

$$u(0, x) = u_0(x), \quad x \in \Omega. \tag{4.34}$$

Here $A(x, D)$ is a second order Waldenfels operator acting on the space of twice differentiable functions defined on Ω and defined as

$$A(x,D)\varphi(x) = c_0(x)\varphi(x) + \sum_{i=1}^{d} b_i(x)\frac{\partial\varphi(x)}{\partial x_i} + \sum_{ij=1}^{d} a_{ij}(x)\frac{\partial^2\varphi(x)}{\partial x_i \partial x_j}$$

$$+ \int_{\Omega}\left[\varphi(y) - \phi(x,y)\left(\varphi(x) - \sum_{i=1}^{d}(y_i - x_i)\frac{\partial\varphi(x)}{\partial y_i}\right)\right]\nu(x,dy), \quad x \in \Omega, \quad (4.35)$$

and $\mathcal{W}(x',D)$ is a boundary pseudo-differential operator acting on the space of twice differentiable functions defined on $\partial\Omega$ through the local coordinates $x' = (x_1,\ldots,x_{n-1}) \in \partial\Omega$ (see [Taira (1992)]):

$$\mathcal{W}(x',D) = Q(x',D) + \mu(x')\frac{\partial}{\partial\mathbf{n}} - \delta(x')A(x',D) + \Gamma_1(x',D) + \Gamma_2(x',D) \quad (4.36)$$

with (pseudo)-differential operators

$$Q(x',D)\varphi(x') = \sum_{j,k=1}^{d-1} \alpha_{jk}(x')\frac{\partial^2\varphi(x')}{\partial x_j \partial x_k} + \sum_{j=1}^{d-1} \beta_j(x')\frac{\partial\varphi(x')}{\partial x_j} + \gamma(x')\varphi(x'), \quad (4.37)$$

$$\Gamma_1(x',D)\varphi(x')$$

$$= \int_{\partial\Omega}\left[\varphi(y') - \tau_1(x',y')\left(\varphi(x') + \sum_{k=1}^{d-1}(y_k - x_k)\frac{\partial\varphi(x')}{\partial x_k}\right)\right]\nu_1(x',dy'), \quad (4.38)$$

and

$$\Gamma_2(x',D)\varphi(x')$$

$$= \int_{\Omega}\left[\varphi(y) - \tau_2(x',y)\left(\varphi(x') + \sum_{k=1}^{d-1}(y_k - x_k)\frac{\partial\varphi(x')}{\partial x_k}\right)\right]\nu_2(x',dy). \quad (4.39)$$

Here $\tau_j(x',y')$, $j = 1,2$, are local unit functions and $\nu_j(x',dy')$, $j = 1,2$, are Lévy kernels satisfying some conditions indicated below, and u_0 is the density function of the initial state X_0. If the initial state $X_0 = 0$, then $u_0(x) = \delta_0(x)$. The boundary condition (4.33) is called a *second order Wentcel's boundary condition*, to credit Wentcel's contribution [Wentcel (1959)] to the theory of diffusion processes.

Let the coefficients of the Waldenfels operator (4.35) as well as the local unit function and Lévy kernel of this operator satisfy the following conditions:

(i) $a_{ij}(x) \in C^{\infty}(\Omega) \cap C(\overline{\Omega})$, $a_{ij}(x) = a_{ji}(x)$ for all $i,j = 1,\ldots,n$, and $x \in \overline{\Omega}$, and there exists a constant $a_0 > 0$ such that

$$\sum_{i,j=1}^{d} a_{ij}(x)\xi_i\xi_j \geqslant a_0|\xi|^2, \quad x \in \overline{\Omega}, \ \xi \in \mathbb{R}^n;$$

(ii) $b_j(x) \in C^{\infty}(\Omega) \cap C(\overline{\Omega})$, $j = 1,\ldots,n$;

(iii) $c_0(x) \in C^{\infty}(\Omega) \cap C(\overline{\Omega})$, and $c_0(x) \leqslant 0$ in $\overline{\Omega}$;

(iv) the local unit function $\phi(x,y)$ and the kernel function $\nu(x,dy)$ are such that the symbol $\sigma_A(x,\xi)$ of the operator $A(x,D)$ belongs to the class of symbols $S(\Omega \times \mathbb{R}^d) = C(\overline{\Omega} \times \mathbb{R}^d) \cap C^{\infty}(\Omega \times (\mathbb{R}^d\backslash G))$, where $G \subset \mathbb{R}^d$ is a set of d-dimensional Lebesgue measure zero; see [Umarov (2015b)] for the theory of pseudo-differential operators with such non-regular symbols.

Equation (4.32) describes a diffusion process accompanied by jumps in Ω with the drift vector $(b_1(x), \ldots, b_d(x))$ and diffusion coefficient defined by the matrix-function $(a_{ij}(x))_{i,j=1,\ldots,d}$, and jumps governed by the Lévy measure $\nu(x, \cdot)$. We assume additionally that the condition

(C1) $\qquad A(x, D)[1(x)] = c_0(x) + \displaystyle\int_{\Omega} [1 - \phi(x, y)]\nu(x, dy) \leqslant 0, \quad x \in \Omega,$

is fulfilled to ensure that the jump phenomenon from $x \in \Omega$ to the outside of a neighborhood of x is "dominated" by the absorption phenomenon at x (see [Taira (1992)] for details).

The coefficients $\alpha_{ij}(x')$, $\beta_j(x')$, $i, j = 1, \ldots, d-1$, and $\gamma(x')$ of the operator $Q(x', D)$ in (4.37) satisfy the following conditions:

(a) $\alpha_{ij}(x') \in C^{\infty}(\partial\Omega)$ and $\alpha_{ij}(x') = \alpha_{ji}(x')$ for all $i, j = 1, \ldots, d-1$, and $x' \in \partial\Omega$, and there exists a constant $\alpha_0 > 0$ such that

$$\sum_{i,j=1}^{d-1} \alpha_{ij}(x')\xi_i\xi_j \geqslant \alpha_0 |\xi|^2, \quad x' \in \partial\Omega, \ \xi \in \mathbb{R}^{d-1};$$

(b) $\beta_j(x') \in C^{\infty}(\partial\Omega)$, $j = 1, \ldots, d-1$;
(c) $\gamma(x') \in C^{\infty}(\partial\Omega)$ and $\gamma(x') \leqslant 0$ in $\partial\Omega$.

The symbols $\sigma_{\Gamma_1}(x', \xi)$ and $\sigma_{\Gamma_2}(x', \xi)$ of the boundary pseudo-differential operators $\Gamma_1(x', D)$ and $\Gamma_2(x', D)$ in equations (4.38) and (4.39) satisfy the following condition:

(d) the local unit functions $\phi_k(x, y)$, $k = 1, 2$, and the kernel functions $\nu_k(x, dy)$, $k = 1, 2$, are such that the symbols $\sigma_{\Gamma_k}(x', \xi)$ of operators $\Gamma_k(x', D)$, $k = 1, 2$, belong to the class of symbols $S(\partial\Omega, \mathbb{R}^d) = C^{\infty}(\partial\Omega \times (\mathbb{R}^d \backslash G_0))$, where $G_0 \in \mathbb{R}^d$ is a set of d-dimensional measure zero.

The boundary condition (4.33) with the operator $\mathcal{W}(x, D)$ defined in equations (4.36)–(4.39) describes a combination of continuous diffusion and jump processes taking place on the boundary, as well as jumps from the boundary into the region, and the viscosity phenomenon near the boundary. Namely, the term

$$\sum_{j,k=1}^{d-1} \alpha_{jk}(x') \frac{\partial^2 u(t, x')}{\partial x_j \partial x_k} + \sum_{j=1}^{d-1} \beta_j(x') \frac{\partial u(t, x')}{\partial x_j}$$

governs the diffusion process on the boundary, the term $\gamma(x')u(t, x')$ is responsible for the absorption phenomenon at $x' \in \partial\Omega$, the term $\mu(x')\frac{\partial}{\partial \mathbf{n}}$ expresses the reflexion phenomenon at $x' \in \partial\Omega$, the term $\delta(x')A(x', D)u(t, x')$ expresses the viscosity near $x' \in \partial\Omega$, and the terms $\Gamma_1(x', D)u(t, x')$ and $\Gamma_2(x', D)u(t, x')$ govern jump processes on the boundary and jump processes from the boundary into the region, respectively. We assume that the condition

(C2) $\mathcal{W}(x', D)[1(x)] = \gamma(x') + \displaystyle\int_{\partial\Omega} [1 - \tau_1(x', y')]\nu_1(x', dy')$

$$+ \int_{\Omega} [1 - \tau_2(x', y)]\nu_2(x', dy) \leqslant 0, \quad x' \in \partial\Omega,$$

is fulfilled to ensure that the jump phenomenon from $x' \in \partial\Omega$ to the outside of a neighborhood of x' on the boundary $\partial\Omega$ or inward to the region Ω is "dominated" by the absorption phenomenon at x'.

We also assume the following transversality condition of the boundary operator $\mathcal{W}$:

(C3) $\displaystyle\int_{\Omega} \nu_2(x', dy) = \infty$ if $\mu(x') = \delta(x') = 0.$

Define the operator $\mathfrak{U}_{\mathcal{W}}$ as

$$\mathfrak{U}_{\mathcal{W}}\phi(x) = A(x, D)\phi(x), \quad x \in \Omega, \tag{4.40}$$

with the domain

$$\mathrm{Dom}(\mathfrak{U}_{\mathcal{W}}) = \{\phi \in C(\bar{\Omega}) : A(x, D)\phi \in C(\bar{\Omega}), \, \mathcal{W}(x', D)\phi(x') = 0 \text{ for } x' \in \partial\Omega\}, \tag{4.41}$$

where $A(x, D)$ is a Waldenfels operator defined in (4.35) and $\mathcal{W}(x', D)$ is Wentcel's boundary operator defined in (4.36).

In Section 4.4 we introduced Feller semigroups associated with stochastic processes in $\mathbb{R}^d$. Feller semigroups in the case of a bounded domain Ω with smooth boundary $\partial\Omega$ are defined as follows.

Definition 4.10. A strongly continuous, positivity preserving, contractive semigroup $\{T_t\}_{t\geqslant 0}$ defined on $C(\bar{\Omega})$ and such that $T_t\left[C(\bar{\Omega})\right] \subset C(\bar{\Omega})$ is called a Feller semigroup on $\bar{\Omega}$.

The following theorem due to Taira [Taira (1992), Taira (2004)] provides general conditions for the operator $\mathfrak{U}_{\mathcal{W}}$ to generate a Feller semigroup.

Theorem 4.5. [Taira (2004), Thm. 1.2] *Let the conditions* $(i) - (iv)$, $(a) - (d)$, *and* $(C1) - (C3)$ *be verified. Then the operator* $\mathfrak{U}_{\mathcal{W}}$ *defined in* (4.40)–(4.41) *generates a Feller semigroup on* $\bar{\Omega}$.

Remark 4.2. We note that Theorem 4.5 requires the transversality condition (C3) on the boundary $\partial\Omega$. Taira [Taira (2004)] generalized this theorem for the non-transversality case of the boundary as well (see [Taira (2004), Thm. 1.3–1.5]).

In Section 7.7, we will use the following theorem on the existence of a unique solution of the initial-boundary value problem for the pseudo-differential equations

$$\frac{\partial u(t, x)}{\partial t} = A(x, D)u(t, x), \quad t > 0, \, x \in \Omega, \tag{4.42}$$

$$\mathcal{W}(x', D)u(t, x') = 0, \quad t > 0, \, x' \in \partial\Omega, \tag{4.43}$$

$$u(0, x) = u_0(x), \quad x \in \Omega. \tag{4.44}$$

To state the theorem, let $C^2_{\mathcal{W}}(\Omega) = \{\varphi \in C^2(\Omega) : \mathcal{W}(x', D)\varphi(x') = 0, \ x' \in \partial\Omega\}$. Also, let $C^1(t > 0; C^2_{\mathcal{W}}(\Omega))$ denote the space of vector functions that are differentiable in t and belonging to $C^2_{\mathcal{W}}(\Omega)$ for each fixed $t > 0$.

Theorem 4.6. *Let the conditions* $(i) - (iv)$, $(a) - (d)$, *and* $(C1) - (C3)$ *be verified. Then initial-boundary value problem* (4.42)–(4.44) *has a unique solution* $u(t, x)$ *in the space* $C([0, \infty) \times \bar{\Omega}) \cap C^1(t > 0; C^2_{\mathcal{W}}(\Omega))$.

Proof. Let the operator $\mathfrak{U}_{\mathcal{W}}$ be defined as in equations (4.40)–(4.41). If the conditions of the theorem are verified, then by Theorem 4.5, there exists a Feller semigroup $\{T_t\}_{t \geqslant 0}$ on $\bar{\Omega}$ generated by the operator $\mathfrak{U}_{\mathcal{W}}$. By Proposition 4.8, for an arbitrary $u_0 \in C(\bar{\Omega})$, the function $u(t, x) = T_t u_0(x)$, $t \geqslant 0$, $x \in \Omega$, exists and solves the following initial value problem for a differential-operator equation

$$\frac{\partial u(t, x)}{\partial t} = \mathfrak{U}_{\mathcal{W}} u(t, x), \quad t \geqslant 0, \ x \in \Omega, \tag{4.45}$$

$$\lim_{t \to 0} u(t, x) = u_0(x), \quad x \in \Omega. \tag{4.46}$$

Moreover, it follows from the general operator theory that the equality

$$u(t, x) = T_t u_0(x) = e^{t \mathfrak{U}_{\mathcal{W}}} u_0(x) \tag{4.47}$$

holds. Since the operator $A(x, D)$ is elliptic and $\mathfrak{U}_{\mathcal{W}}$ has a spectrum in the negative real axis, it follows from the smoothness of a solution to a parabolic equation that $u(t, x)$ has all derivatives at $t > 0$ and $x \in \Omega$. Thus, in particular, this function belongs to the space $C([0, \infty) \times \bar{\Omega}) \cap C^1(t > 0; C^2_{\mathcal{W}}(\Omega))$. The uniqueness of the solution is implied by equation (4.47). $\qquad\square$

Chapter 5

Stochastic processes and time-changes

Introduction

This chapter introduces the Skorokhod space and various types of stochastic processes that are needed for discussions in Chapters 6 and 7. Section 5.1 studies the Skorokhod space and its relevant topologies, which are indispensable for descriptions of scaling limits of continuous time random walks where the limits are given by time-changed stochastic processes (see Sections 6.3 and 6.5). In Section 5.2, semimartingales are presented as processes for which the Itô-type stochastic integration is valid and general discussions of time-changes are also provided. Lévy processes, which form an important subclass of semimartingales, are discussed in Sections 5.3 and 5.4, with an emphasis on stable processes and stable subordinators. Section 5.5 is devoted to Gaussian processes, which include examples of non-semimartingales. For general Gaussian processes, the Itô-type stochastic integrals cannot be defined. Brownian motion provides a special case of both Lévy and Gaussian processes, but in general, Lévy processes have properties that are very different from those of Gaussian processes.

5.1   The Skorokhod space and its relevant topologies

A function $x : [0, \infty) \to \mathbb{R}^d$ is said to be *càdlàg* if x is right-continuous at any $t \geq 0$, i.e. $\lim_{s \to t+} x(s) = x(t)$, and has left limits at any $t > 0$, i.e. $\lim_{s \to t-} x(s) = x(t-)$ exists. A function $x : [0, \infty) \to \mathbb{R}^d$ is said to be *càglàd* if x is left-continuous at any $t > 0$, i.e. $\lim_{s \to t-} x(s) = x(t)$, and has right limits at any $t \geq 0$, i.e. $\lim_{s \to t+} x(s) = x(t+)$ exists. For convenience, this section considers only càdlàg functions and uses the convention $x(0-) = 0$. The space of all càdlàg functions defined on $[0, \infty)$ is denoted by $\mathbb{D}([0, \infty), \mathbb{R}^d)$ and called the *Skorokhod space*. The Skorokhod space $\mathbb{D}([0, \infty), \mathbb{R}^d)$ is complete with respect to the locally uniform topology generated by the metric

$$d(x, y) = \sum_{n=1}^{\infty} \frac{\min(1, \|x - y\|_n)}{2^n},$$

where

$$\|x\|_n = \sup_{0 \leqslant t \leqslant n} |x(t)| = \sup_{0 \leqslant t \leqslant n} (|x_1(t)|^2 + \cdots + |x_d(t)|^2)^{1/2}.$$

However, $\mathbb{D}([0, \infty), \mathbb{R}^d)$ is not separable with respect to this topology. Moreover, even though the uniform topology works well in the subspace $C([0, \infty), \mathbb{R}^d)$ of continuous functions, it fails to be efficient in approximations of functions in $\mathbb{D}([0, \infty), \mathbb{R}^d)$.

Below we provide some facts related to the Skorokhod space with non-uniform topologies. There are two topologies on the Skorokhod space that are useful and frequently used: the J_1 topology and the M_1 topology. These topologies were introduced by A.V. Skorokhod in 1956 (see [Skorokhod (1956)]). To define the J_1 topology we introduce the set Λ of continuous strictly increasing functions λ on $[0, \infty)$ such that $\lambda(0) = 0$ and $\lim_{t \to \infty} \lambda(t) = \infty$. The J_1 *topology* is defined by the metric

$$\delta(x, y) = \sum_{n=1}^{\infty} \frac{\min(1, \omega_n(x, y))}{2^n}, \tag{5.1}$$

where

$$\omega_n(x, y) = \inf_{\lambda \in \Lambda} (\|\lambda - I\|_n + \|x - y \circ \lambda\|_n)$$

with I being the identity function. The Skorokhod space endowed with the J_1 topology is denoted by $\mathbb{D}([0, \infty), \mathbb{R}^d, J_1)$.

The space $\mathbb{D}([0, \infty), \mathbb{R}^d, J_1)$ can be defined as a projective limit of a sequence of the Skorokhod spaces defined on finite intervals. This approach works for other topologies as well. Let $\{t_n\}$ be an increasing sequence of positive numbers:

$$0 < t_1 < t_2 < \cdots \quad \text{with} \quad t_n \to \infty \quad \text{as} \quad n \to \infty. \tag{5.2}$$

Let $\mathbb{D}_n = \mathbb{D}([0, t_n), \mathbb{R}^d, \tau_n)$ be the space of càdlàg functions on $[0, t_n)$ endowed with a topology τ_n, where τ_{n+1} is weaker than τ_n (that is, $\tau_{n+1} \prec \tau_n$) for all $n \geqslant 1$. Since the sequence of spaces $\{\mathbb{D}_n; n \geqslant 1\}$ satisfies the condition

$$\mathbb{D}_1 \supset \mathbb{D}_2 \supset \cdots \supset \mathbb{D}_n \supset \mathbb{D}_{n+1} \supset \cdots \tag{5.3}$$

and the sequence of topologies $\{\tau_n; n \geqslant 1\}$ satisfies the condition

$$\tau_1 \succ \tau_2 \succ \cdots \succ \tau_n \succ \tau_{n+1} \succ \cdots, \tag{5.4}$$

the space $\mathbb{D}_{n+1}$ is continuously embedded to the space $\mathbb{D}_n$ for each $n \geqslant 1$. Hence, we can define the projective limit of $\mathbb{D}_n$:

$$\mathbb{D}_\infty(\tau) = pr \lim_{n \to \infty} \mathbb{D}_n := \cap_{n=1}^{\infty} \mathbb{D}_n,$$

which is endowed with the coarsest topology $\tau = \lim_{n \to \infty} \tau_n$. If the topology τ_n is induced by the J_1-metric

$$\omega_n'(x, y) = \inf_{\lambda \in \Lambda_n} (\|\lambda - I\|_{t_n} + \|x - y \circ \lambda\|_{t_n}),$$

where Λ_n is the set of continuous strictly increasing functions mapping $[0, t_n]$ onto $[0, t_n]$ and $\|a\|_{t_n} = \sup_{0 \leqslant t \leqslant t_n} |a(t)|$, then the sequence $\{\tau_n\}$ satisfies (5.4). The coarsest topology in this case is the topology induced by the metric

$$\delta'(x, y) = \sum_{n=1}^{\infty} \frac{\min(1, \omega_n'(x, y))}{2^n}$$

and is equivalent to the J_1 topology induced by the metric $\delta(x, y)$ defined in (5.1). Hence, in this case,

$$\mathbb{D}_\infty(\tau) = \mathbb{D}([0, \infty), \mathbb{R}^d, J_1).$$

To define the M_1 topology, let $\{t_n\}$ be a sequence satisfying (5.2). We first introduce the notion of the completed graph of a function $x \in \mathbb{D}([0, t_n), \mathbb{R}^d)$, which takes into account straight lines connecting $(t, x(t-))$ with $(t, x(t))$ in the cross sections $\{t\} \times \mathbb{R}^d$ of the space $[0, t_n) \times \mathbb{R}^d$ with the discontinuity points $t \in [0, t_n)$. Namely, the *completed graph* Γ_x of $x \in \mathbb{D}([0, t_n), \mathbb{R}^d)$ is defined as

$$\Gamma_x = \{(t, z) \in [0, t_n) \times \mathbb{R}^d : z = \alpha x(t-) + (1 - \alpha)x(t) \text{ for some } \alpha \in [0, 1]\}.$$

One can define an order relation in Γ_x. We say that $(t_1, z_1) < (t_2, z_2)$ if either $t_1 < t_2$, or $t_1 = t_2$ and $|x(t_1-) - z_1| < |x(t_2-) - z_2|$. By a *parametric representation* of the function x, we mean a continuous nondecreasing function (r, u) mapping $[0, t_n)$ onto Γ_x, where r is the time component and u is the spatial component of the completed graph Γ_x. Denote by Π_x the set of parametric representations of $x \in \mathbb{D}([0, t_n), \mathbb{R}^d)$. The M_1 *topology* on $\mathbb{D}([0, t_n), \mathbb{R}^d)$ is induced by the metric

$$\rho_n(x, y) = \inf_{\substack{(r,u)\in\Pi_x \\ (s,v)\in\Pi_y}} \max\{\|r - s\|_{t_n}, \|u - v\|_{t_n}\}.$$

The fact that $\rho_n(x, y)$ indeed is a metric is found in [Whitt (2002)]. Moreover, it is easy to verify that the topologies induced by $\rho_n(x, y)$ satisfy condition (5.4). Hence, we can define

$$\mathbb{D}([0, \infty), \mathbb{R}^d, M_1) := pr \lim_{n \to \infty} \mathbb{D}([0, t_n), \mathbb{R}^d, \rho_n) \tag{5.5}$$

with the coarsest topology of the projective limit, which is equivalent to the topology induced by the metric

$$m(x, y) = \sum_{n=1}^{\infty} \frac{\min(1, \rho_n(x, y))}{2^n}.$$

Further, consider a sequence of continuous functions κ_n, $n = 1, 2, \ldots$, given by

$$\kappa_n(t) = \begin{cases} 1, & \text{if } t < t_n, \\ \frac{t_{n+1}-t}{t_{n+1}-t_n}, & \text{if } t_n \leqslant t < t_{n+1}, \\ 0, & \text{if } t \geqslant t_{n+1}, \end{cases}$$

and define the following metric in $\mathbb{D}([0, t_{n+1}), \mathbb{R}^d)$:

$$\pi_n(x, y) = \inf_{\lambda \in \Lambda_{n+1}} (\||\lambda\||_n + \|\kappa_n x - (\kappa_n y) \circ \lambda\|),$$

where

$$\|\lambda\|_n = \sup_{0 \leqslant s < t \leqslant t_{n+1}} \left| \log \frac{\lambda(t) - \lambda(s)}{t - s} \right|.$$

The fact that $\pi_n(x, y)$ is a metric is found in [Jacod and Shiryaev (1987)]. It is easy to verify that the topologies induced by $\pi_n(x, y)$ satisfy condition (5.4). Hence, we can define

$$\mathbb{D}([0, \infty), \mathbb{R}^d, \pi) := pr \lim_{n \to \infty} \mathbb{D}([0, t_n), \mathbb{R}^d, \pi_n) \tag{5.6}$$

with the coarsest topology of the projective limit, which is equivalent to the topology induced by the metric

$$\pi(x, y) = \sum_{n=1}^{\infty} \frac{\min(1, \pi_n(x, y))}{2^n}.$$

We note that the Skorokhod space is not complete under the J_1 and M_1 topologies. However, as the following proposition states, it is complete under the π-topology.

Proposition 5.1. [Jacod and Shiryaev (1987)] *The space $\mathbb{D}([0, \infty), \mathbb{R}^d)$ is complete under the topology π.*

A sequence $\{x_k\} \subset \mathbb{D}([0, \infty), \mathbb{R}^d)$ converges to $x \in \mathbb{D}([0, \infty), \mathbb{R}^d)$ in one of the topologies if $\{x_k\}$ converges to x in a metric defining the corresponding topology. The projective limit structure used above for the definition of the Skorokhod space of càdlàg functions on the positive real line $[0, \infty)$ allows a characterization of the convergence via the corresponding convergence on finite intervals. For instance, a sequence converges in $\mathbb{D}([0, \infty), \mathbb{R}^d)$ in the J_1 topology if its restriction to $\mathbb{D}([0, t_n), \mathbb{R}^d)$ converges in the J_1 topology for each n.

The following statement establishes a relation between the J_1 and M_1 topologies.

Proposition 5.2. [Whitt (2002)] *For all x_1, $x_2 \in \mathbb{D}([0, \infty), \mathbb{R}^d)$,*

$$m(x_1, x_2) \leqslant \delta(x_1, x_2). \tag{5.7}$$

Thus, any sequence in $\mathbb{D}([0, \infty), \mathbb{R}^d)$ convergent in the J_1 topology is convergent in the M_1 topology as well.

Further properties of the Skorokhod space $\mathbb{D}([0, \infty), \mathbb{R}^d)$ are collected in the next proposition. Let $Disc(x)$ be the set of discontinuity points of $x \in \mathbb{D}([0, \infty), \mathbb{R}^d)$; namely, $Disc(x) = \{t \in [0, \infty) : x(t-) \neq x(t)\}$. The notations $\xrightarrow{J_1}$ and $\xrightarrow{M_1}$ are used for convergences in the J_1 and M_1 topologies, respectively.

Proposition 5.3.

(1) The set $Disc(x)$ for $x \in \mathbb{D}([0, \infty), \mathbb{R}^d)$ is at most countable.

(2) Let τ be either the J_1 or M_1 topology. Suppose that $x_k \xrightarrow{\tau} x$ in $\mathbb{D}([0, \infty), \mathbb{R}^{d_1}, \tau)$ and $y_k \xrightarrow{\tau} y$ in $\mathbb{D}([0, \infty), \mathbb{R}^{d_2}, \tau)$. If $Disc(x) \cap Disc(y) = \emptyset$, then $(x_k, y_k) \xrightarrow{\tau} (x, y)$ in $\mathbb{D}([0, \infty), \mathbb{R}^{d_1 + d_2}, \tau)$.

(3) Let $x_k \xrightarrow{M_1} x$ in $\mathbb{D}([0, \infty), \mathbb{R}^d, M_1)$ and $y_k \xrightarrow{M_1} y$ in $\mathbb{D}([0, \infty), \mathbb{R}^d, M_1)$. If $Disc(x) \cap Disc(y) = \varnothing$, then $x_k - y_k \xrightarrow{M_1} x - y$ in $\mathbb{D}([0, \infty), \mathbb{R}^d, M_1)$.

(4) For any nondecreasing function $\lambda \in \mathbb{D}([0, \infty), [0, \infty))$ and any $x \in \mathbb{D}([0, \infty), \mathbb{R}^d)$, the function $x \circ \lambda \in \mathbb{D}([0, \infty), \mathbb{R}^d)$.

For proofs and other details, we refer the reader to books [Billingsley (1999), Jacod and Shiryaev (1987), Whitt (2002), Silvestrov (2004)].

We now turn to convergence of stochastic processes in the Skorokhod space. A stochastic process $Z = (Z_t)_{t \geq 0}$ is said to be *càdlàg* (resp. *càglàd*) if Z has right-continuous sample paths with left limits (resp. left-continuous sample paths with right limits). It follows from Proposition 5.3 that the assumption that Z is càdlàg or càglàd requires its sample paths to have at most countably many finite jumps. Associated to a càdlàg process Z is its jump process $(\Delta Z_t)_{t \geq 0}$ where $\Delta Z_t := Z_t - Z_{t-}$ with Z_{t-} denoting the left limit at t and $Z_{0-} = 0$ by convention.

A family of d-dimensional càdlàg stochastic processes $X_\varepsilon = (X_\varepsilon(t))_{t \geq 0}$, $\varepsilon > 0$, is said to converge weakly (in law, in distribution) to a càdlàg stochastic process $X_0 = (X_0(t))_{t \geq 0}$ as $\varepsilon \to 0$ if

$$\mathbb{E}[F(X_\varepsilon)] \to \mathbb{E}[F(X_0)] \quad \text{as} \quad \varepsilon \to 0$$

for all bounded continuous functionals $F : \mathbb{D}([0, \infty), \mathbb{R}^d) \to \mathbb{R}$. Here the continuity of F means that if $x_k \to x$ in the Skorokhod space $\mathbb{D}([0, \infty), \mathbb{R}^d)$ with a certain topology, then $F(x_k) \to F(x)$ in $\mathbb{R}$. In particular, we write $X_\varepsilon \xrightarrow{J_1} X_0$ (or $X_\varepsilon \xrightarrow{M_1} X_0$) if the weak convergence occurs with the J_1 (or M_1) topology.

Convergence of a sequence of stochastic processes in the Skorokhod space is characterized by recognizing the limiting process via weak convergence and compactness of the sequence. These characterizations were given by Skorokhod [Skorokhod (1956)]. Below we present criteria for the J_1 and M_1 convergences given in [Silvestrov (2004)].

For càdlàg stochastic processes X_ε, $\varepsilon > 0$, and X_0, consider the following two conditions:

(C) weak convergence: $(X_\varepsilon(t))_{t \in S}$ converges weakly to $(X_0(t))_{t \in S}$ as $\varepsilon \to 0$, where S is a dense subset of $[0, \infty)$ containing the point 0;

(K_J) J_1-compactness:

$$\lim_{c \to 0} \limsup_{\varepsilon \to 0} \mathbb{P}\{\omega_J(X_\varepsilon(\cdot), c, T) > \delta\} = 0 \quad \text{for all} \quad \delta, T > 0,$$

where

$$\omega_J(x(\cdot), c, T) = \sup_{t-c \leq t_1 < t < t_2 \leq (t+c) \wedge T} \min(|x(t_1) - x(t)|, |x(t_2) - x(t)|).$$

Theorem 5.1. [Silvestrov (2004)] *Conditions (C) and (K_J) are necessary and sufficient for the J_1 convergence $X_\varepsilon \xrightarrow{J_1} X_0$ in $\mathbb{D}([0, \infty), \mathbb{R}^d)$ as $\varepsilon \to 0$.*

For the M_1 convergence, the J_1-compactness condition (K_J) needs to be replaced by the following condition:

(K_M) M_1-compactness:

$$\lim_{c \to 0} \limsup_{\varepsilon \to 0} \ \mathbb{P}\{\omega_M(X_\varepsilon(\cdot), c, T) > \delta\} = 0 \quad \text{for all} \ \ \delta, T > 0,$$

where

$$\omega_M(x(\cdot), c, T) = \sup_{t-c \leqslant t_1 < t_2 < t_3 \leqslant (t+c) \wedge T} \left| x(t_2) - [x(t_1), x(t_3)] \right|$$

with $[x(t_1), x(t_3)] = \{ax(t_1) + (1-a)x(t_3) : \ 0 \leqslant a \leqslant 1\}$.

Theorem 5.2. [Silvestrov (2004)] *Conditions (C) and (K_M) are necessary and sufficient for the M_1 convergence $X_\varepsilon \xrightarrow{M_1} X_0$ in $\mathbb{D}([0, \infty), \mathbb{R}^d)$ as $\varepsilon \to 0$.*

5.2   Semimartingales and time-changes

In the remainder of this chapter, a complete filtered probability space $(\Omega, \mathcal{F}, (\mathcal{F}_t), \mathbb{P})$ is fixed, where the filtration $(\mathcal{F}_t)$ satisfies the *usual conditions*; that is, it is right-continuous and contains all $\mathbb{P}$-null sets in $\mathcal{F}$. Random vectors are assumed to be defined on the probability space and the expectation of a random vector X is denoted by $\mathbb{E}[X]$.

In discussions of Itô-type stochastic integrals, semimartingales play an important role. A càdlàg process Z is called an $(\mathcal{F}_t)$-*semimartingale* if there exist an $(\mathcal{F}_t)$-local martingale M and an $(\mathcal{F}_t)$-adapted process A of finite variation on compact sets such that

$$Z_t = M_t + A_t \quad \text{for} \ \ t \geqslant 0. \tag{5.8}$$

Although this decomposition is not unique in general, the local martingale part M can be uniquely decomposed as $M_t = M_t^c + M_t^d$ with a continuous local martingale M^c and a purely discontinuous local martingale M^d. The process M^c is determined independently of the initial decomposition of Z into M and A, and we write $Z^c := M^c$. (see [Jacod and Shiryaev (1987)], I. Prop. 4.27]).

The class of semimartingales forms a real vector space which is closed under multiplication. It is known to be the largest class of processes for which the Itô-type stochastic integrals are defined as in Chapter 2. Let $\mathcal{P}(\mathcal{F}_t)$ be the smallest σ-algebra on $[0, \infty) \times \Omega$ which makes all càglàd, $(\mathcal{F}_t)$-adapted processes measurable. Given an $(\mathcal{F}_t)$-semimartingale Z, let $L(Z_t, \mathcal{F}_t)$ denote the class of $\mathcal{P}(\mathcal{F}_t)$-measurable, or $(\mathcal{F}_t)$-*predictable*, processes H for which a stochastic integral driven by Z, denoted

$$(H \bullet Z)_t = \int_0^t H_s dZ_s, \tag{5.9}$$

can be constructed. One important property of the Itô-type stochastic integrals is that $H \bullet Z$ is again an $(\mathcal{F}_t)$-semimartingale. For details of the construction of stochastic integrals, consult e.g. [Protter (2004), II–IV].

The *quadratic variation* $[Z, Z] = ([Z, Z]_t)_{t \geqslant 0}$ of a one-dimensional semimartingale Z, which is defined via (2.1), is connected with a stochastic integral via the expression

$$[Z, Z]_t = Z_t^2 - 2 \int_0^t Z_{s-} dZ_s. \tag{5.10}$$

The quadratic variation $[Z, Z]$ is a càdlàg, $(\mathcal{F}_t)$-adapted, nondecreasing process. By polarization, the map $[\cdot, \cdot]$ becomes a symmetric, bilinear form on the class of one-dimensional semimartingales. For semimartingales Y and Z, the notation $[Y, Z]^c$ denotes the continuous part of $[Y, Z]$; namely,

$$[Y, Z]_t^c := [Y, Z]_t - \sum_{0 < s \leqslant t} \Delta[Y, Z]_s = [Y, Z]_t - \sum_{0 < s \leqslant t} \Delta Y_s \cdot \Delta Z_s,$$

where $\Delta X_s = X_s - X_{s-}$ for a given càdlàg process X. It follows by comparing this definition with [Jacod and Shiryaev (1987), I. Thm. 4.52] that

$$[Z, Z]^c = [Z^c, Z^c], \tag{5.11}$$

where Z^c denotes the continuous local martingale part of Z. The following basic properties of stochastic integrals are frequently employed in this section.

Lemma 5.1. *Let* Y *and* Z *be* $(\mathcal{F}_t)$-*semimartingales. Let* $H \in L(Z_t, \mathcal{F}_t)$ *and* $K \in L(Y_t, \mathcal{F}_t)$.

(1) $\Delta(H \bullet Z) = H \cdot \Delta Z$. *In particular, a stochastic integral driven by a continuous semimartingale is again a continuous semimartingale.*
(2) $H \cdot K \in L([Z, Y]_t, \mathcal{F}_t)$ *and* $[H \bullet Z, K \bullet Y] = (H \cdot K) \bullet [Z, Y]$.

Among the most important results in the theory of stochastic integration is the celebrated Itô formula (see e.g. [Protter (2004)]), which establishes a stochastic calculus for stochastic integrals driven by a semimartingale.

Proposition 5.4 (Itô formula). *Let* $X = (X^1, \ldots, X^d)$ *be a* d-*dimensional semimartingale. If* $f \in C^2(\mathbb{R}^d)$, *then* $f(X)$ *is a one-dimensional semimartingale, and, for all* $t \geqslant 0$, *with probability one,*

$$f(X_t) = f(X_0) + \sum_{i=1}^d \int_0^t \frac{\partial f}{\partial x^i}(X_{s-}) dX_s^i \tag{5.12}$$

$$+ \frac{1}{2} \sum_{i,j=1}^d \int_0^t \frac{\partial^2 f}{\partial x^i \partial x^j}(X_{s-}) d[X^i, X^j]_s^c$$

$$+ \sum_{0 < s \leqslant t} \left\{ f(X_s) - f(X_{s-}) - \sum_{i=1}^d \frac{\partial f}{\partial x^i}(X_{s-}) \Delta X_s^i \right\}.$$

One useful implication of the Itô formula (5.12) is the product rule. Namely, if Y and Z are both one-dimensional semimartingales starting at 0, then, for all $t \geqslant 0$, with probability one,

$$Y_t Z_t = \int_0^t Y_{s-} dZ_s + \int_0^t Z_{s-} dY_s + [Y, Z]_t. \tag{5.13}$$

Note that letting $Y = Z$ in (5.13) yields the expression (5.10). These formulas are indispensable tools for working with stochastic differential equations.

A random time $T : \Omega \to [0, \infty]$ is called an $(\mathcal{F}_t)$-*stopping time* if $\{T \leqslant t\} \in \mathcal{F}_t$ for all $t \geqslant 0$. Associated with a given $(\mathcal{F}_t)$-stopping time T is the σ-algebra defined by

$$\mathcal{F}_T = \{C \in \mathcal{F}; \ C \cap \{T \leqslant t\} \in \mathcal{F}_t \text{ for all } t \geqslant 0\}.$$

A càdlàg process $(E_t)_{t \geqslant 0}$ is called an $(\mathcal{F}_t)$-*time-change* if it is a family of $(\mathcal{F}_t)$-stopping times such that the mapping $t \to E_t$ is nondecreasing a.s. It is said to be *finite* if for each $t \geqslant 0$, E_t is finite a.s. For a finite $(\mathcal{F}_t)$-time-change (E_t), the family $(\mathcal{F}_{E_t})_{t \geqslant 0}$ defines a new filtration. It satisfies the usual conditions since the right-continuity of $(\mathcal{F}_t)$ and (E_t) implies that of $(\mathcal{F}_{E_t})$. In addition, for any $(\mathcal{F}_t)$-adapted process Z, the time-changed process (Z_{E_t}) is known to be $(\mathcal{F}_{E_t})$-adapted. In fact, more can be said.

Lemma 5.2. [Jacod (1979), Cor. 10.12] *Let Z be an $(\mathcal{F}_t)$-semimartingale. Let (E_t) be a finite $(\mathcal{F}_t)$-time-change. Then (Z_{E_t}) is an $(\mathcal{F}_{E_t})$-semimartingale.*

Namely, *every time-changed semimartingale is a semimartingale*, and hence, it can serve as an integrator of the Itô-type stochastic integrals (however, we should note that the filtration must be also time-changed). On the other hand, the local martingale property is not always preserved under a finite time-change. A simple example is an $(\mathcal{F}_t)$-adapted Brownian motion $Z = B$ with the finite $(\mathcal{F}_t)$-time-change (E_t) defined by $E_t = \inf\{s > 0; B_s = t\}$. For each fixed t, since E_t is finite a.s. and represents the first time the Brownian motion B hits the level t, it follows that $B_{E_t} = t$. Hence, with this specific time-change (E_t), the time-changed Brownian motion (B_{E_t}) is a deterministic process of finite variation and is no longer a local martingale. One way to exclude this unexpected possibility is to introduce the notion of synchronization. A process Z is said to be *in synchronization* with a time-change E if Z is constant on every interval of the form $[E_{t-}, E_t]$ a.s. Other properties that a time-change preserves appear in [Jacod (1979), Thm. 10.16]. In the literature, books [Jacod (1979), Kallsen and Shiryaev (2002)] use the expression "adapted" in describing a process being in synchronization with a time-change. A different terminology "continuous" is used in [Revuz and Yor (1999)]. Nevertheless, the phrase "in synchronization" is adopted here to avoid any possible confusions or misunderstandings that the other expressions may create.

In Section 6.1, the concept of synchronization is further investigated; it turns out to be an essential concept in developing stochastic calculus for stochastic integrals driven by time-changed semimartingales.

5.3 Lévy processes

Lévy processes form an important subclass of semimartingales and are widely used in many applied areas. A càdlàg process $L = (L_t)_{t \geq 0}$ in $\mathbb{R}^n$ is called a *Lévy process* if L has independent and stationary increments and the mapping $t \to L_t$ is continuous in probability; i.e. $\lim_{s \to t} \mathbb{P}(|L_t - L_s| > \varepsilon) = 0$ for all $\varepsilon > 0$ and $t \geq 0$. Note that the continuity in probability does not imply that sample paths are continuous; in fact, sample paths of many important Lévy processes have jumps. The simplest examples of Lévy processes include Brownian motion and Poisson processes, and the sum of independent Lévy processes is again a Lévy process.

Lévy processes are characterized by three parameters: a vector $b_0 \in \mathbb{R}^n$, a non-negative definite $n \times n$ matrix Σ, and a measure ν defined on $\mathbb{R}^n \backslash \{0\}$ such that $\int (1 \wedge |w|^2) \nu(dw) < \infty$, called the Lévy measure, where $a \wedge b = \min\{a, b\}$. The *Lévy–Khintchine formula* characterizes the Lévy process L as an infinitely divisible process. Namely, the characteristic function for L is given by $\mathbb{E}[e^{i(\xi, L_t)}] = e^{t\Psi(\xi)}$, where

$$\Psi(\xi) = i(b_0, \xi) - \frac{1}{2}(\Sigma \xi, \xi) + \int_{\mathbb{R}^n \backslash \{0\}} (e^{i(w, \xi)} - 1 - i(w, \xi) \boldsymbol{I}_{(-1,1)}(w)) \nu(dw). \quad (5.14)$$

The function Ψ is called the *Lévy symbol* of L. The Lévy symbol Ψ is continuous, hermitian, conditionally positive definite and $\Psi(0) = 0$ (see e.g. [Applebaum (2009), Thm 1.2.17]). Notice that if $b_0 = 0$, $\Sigma = I$ (the identity matrix) and $\nu \equiv 0$ in (5.14), then L becomes n-dimensional Brownian motion.

The Lévy–Itô decomposition states that a given Lévy process L is represented as

$$L_t = \tilde{b}_0 t + \sigma_0 B_t + \int_{|w|<1} w \tilde{N}(t, dw) + \int_{|w| \geq 1} w N(t, dw), \quad (5.15)$$

where $\tilde{b}_0 \in \mathbb{R}^n$, σ_0 is an $n \times m$-matrix, B is an m-dimensional Brownian motion, and N and $\tilde{N}$ are a Poisson random measure and a compensated Poisson martingale-valued measure on $[0, \infty) \times (\mathbb{R}^n \backslash \{0\})$, respectively (see [Applebaum (2009), Barndorf-Nielsen et al. (2001), Sato (1999)]). Namely, $N(t, A)$ represents the number of jumps of size A up to time t, and $(\int_{|w| \geq 1} w N(t, dw))_{t \geq 0}$ is a compound Poisson process describing large jumps, whereas $(\int_{|w|<1} w \tilde{N}(t, dw))_{t \geq 0}$ is the compensated sum of small jumps. The matrices σ_0 and Σ in equations (5.15) and (5.14), respectively, are related as $\Sigma = \sigma_0 \times \sigma_0^T$, where σ_0^T is the transpose of σ_0. Vectors b_0 and $\tilde{b}_0$ responsible for the drift are not necessarily the same. The Lévy–Itô decomposition implies that the Lévy process L is a semimartingale, that is, $L_t = M_t + A_t$, where

$$M_t = \sigma_0 B_t + \int_{|w|<1} w \tilde{N}(t, dw)$$

is a local martingale and

$$A_t = \tilde{b}_0 t + \int_{|w| \geq 1} w N(t, dw)$$

is a finite variation process. Therefore, stochastic differential equations (SDEs) driven by L are understood within the framework of the Itô-type stochastic integrals discussed in the previous section. A general form of such SDEs is given by

$$Y_t = x + \int_0^t b(Y_{s-})ds + \int_0^t \sigma(Y_{s-})dB_s \qquad (5.16)$$

$$+ \int_0^t \int_{|w|<1} H(Y_{s-}, w)\tilde{N}(ds, dw) + \int_0^t \int_{|w|\geq 1} K(Y_{s-}, w)N(ds, dw),$$

or in shorthand,

$$dY_t = b(Y_{t-})dt + \sigma(Y_{t-})dB_t \qquad (5.17)$$

$$+ \int_{|w|<1} H(Y_{t-}, w)\tilde{N}(dt, dw) + \int_{|w|\geq 1} K(Y_{t-}, w)N(dt, dw)$$

with $Y_0 = x \in \mathbb{R}^n$. Here the continuous mappings $b : \mathbb{R}^n \to \mathbb{R}^n$, $\sigma : \mathbb{R}^n \to \mathbb{R}^{n\times m}$, $H : \mathbb{R}^n \times \mathbb{R}^n \to \mathbb{R}^n$, and $K : \mathbb{R}^n \times \mathbb{R}^n \to \mathbb{R}^n$ satisfy the Lipschitz and growth conditions. Namely, there exist positive constants C_1 and C_2 satisfying

$$|b(y_1) - b(y_2)|^2 + \|\sigma(y_1) - \sigma(y_2)\|^2 + \int_{|w|<1} |H(y_1, w) - H(y_2, w)|^2 \nu(dw) \quad (5.18)$$

$$\leq C_1 |y_1 - y_2|^2 \quad \text{for all } y_1, y_2 \in \mathbb{R}^n;$$

$$\int_{|w|<1} |H(y, w)|^2 \nu(dw) \leq C_2(1 + |y|^2) \quad \text{for all } y \in \mathbb{R}^n. \qquad (5.19)$$

Under these conditions, SDE (5.16) has a unique strong solution Y (see [Applebaum (2009), Situ (2005)]). If the coefficients b, σ, H, and K are bounded, then

$$(T_t\varphi)(x) := \mathbb{E}[\varphi(Y_t)|Y_0 = x]$$

defines a strongly continuous contraction semigroup $\{T_t\}$ defined on the Banach space $C_0(\mathbb{R}^n)$, the space of continuous functions vanishing at infinity. Moreover, the function $u(t, x) = (T_t\varphi)(x)$ satisfies the pseudo-differential equation

$$\frac{\partial}{\partial t}u(t, x) = \mathcal{A}(x, D_x)u(t, x), \quad t > 0, \ x \in \mathbb{R}^n, \qquad (5.20)$$

where the infinitesimal generator $\mathcal{A}(x, D_x)$ is a pseudo-differential operator with the symbol

$$\Psi(x, \xi) = i\langle b(x), \xi\rangle - \frac{1}{2}\langle \Sigma(x)\xi, \xi\rangle \qquad (5.21)$$

$$+ \int_{\mathbb{R}^n\setminus\{0\}} (e^{i\langle G(x,w),\xi\rangle} - 1 - i\langle G(x, w), \xi\rangle \boldsymbol{I}_{(-1,1)}(w))\,\nu(dw).$$

Here $G(x, w) = H(x, w)$ if $|w| < 1$, and $G(x, w) = K(x, w)$ if $|w| \geq 1$ (see [Applebaum (2009), Situ (2005)]). For each fixed $x \in \mathbb{R}^n$, the symbol $\Psi(x, \xi)$ is continuous, hermitian, and conditionally positive definite (see [Courrége (1964), Jacob

(2001, 2002, 2005)]). Using $D_x = -i(\partial_{x^1}, \ldots, \partial_{x^n})$, the pseudo-differential operator $\mathcal{A}(x, D_x)$ can be written as

$$\mathcal{A}(x, D_x)\varphi(x) = i(b(x), D_x)\varphi(x) - \frac{1}{2}(\Sigma(x)D_x, D_x)\varphi(x) \qquad (5.22)$$

$$+ \int_{\mathbb{R}^n \setminus \{0\}} [\varphi(x + G(x, w)) - \varphi(x) - i\boldsymbol{I}_{(-1,1)}(w)(G(x, w), D_x)\varphi(x)]\nu(dw).$$

Here, the domain of $\mathcal{A}(x, D_x)$ contains $C_0^2(\mathbb{R}^n)$. The pseudo-differential equation (5.20) is the backward Fokker–Planck–Kolmogorov (FPK) equation associated with the solution Y to SDE (5.17); hence, there is a connection between the class of SDEs of the form (5.17) and the class of pseudo-differential equations of the form (5.20). Moreover, as in Section 2.4, the Itô formula provides a connection between SDEs of the form (5.17) and the associated forward FPK equations.

Stable Lévy processes are among the most important Lévy processes both theoretically and practically. To discuss them, we first define a stable random variable. A random variable X is called *stable* if there exist sequences (c_n) and (d_n) with $c_n > 0$ such that

$$\sum_{i=1}^n X_i \stackrel{\mathrm{d}}{=} c_n X + d_n,$$

where $X_1, \ldots, X_n$ are independent copies of X and the notation $\stackrel{\mathrm{d}}{=}$ means equality in distribution. In particular, X is called *strictly stable* if $d_n = 0$ for all n. It is known that the only possible choice for c_n is $\sigma n^{1/\alpha}$ for some $\alpha \in (0, 2]$ and $\sigma > 0$. The constant α is called the *stability index*, or simply *index*, of the stable law. The case when $\alpha = 2$ corresponds to a normal distribution, while when $\alpha \neq 2$, the density shows a slower, polynomial decay. In particular, unless $\alpha = 2$, stable random variables do not possess a finite second moment. Moreover, they have a finite first absolute moment if and only if $\alpha \in (1, 2]$. Notice also that closed forms of stable densities are known only in some special cases. (see e.g. [Applebaum (2009), Sato (1999)] for details.) A stable law with stability index α is sometimes referred to as an α-stable law.

Note that the definition of stable laws implies that appropriately scaled random walks of i.i.d. (independent and identically distributed) α-stable random variables

$$\frac{1}{\sigma n^{1/\alpha}}\left(\sum_{i=1}^n X_i - d_n\right)$$

necessarily have the same α-stable law. More generally, there are random walks whose scaling limits possess an α-stable (or strictly α-stable) law. Such random variables are said to belong to the *domain (or strict domain) of attraction of a stable law with index α*. In particular, if i.i.d. random variables $X_1, X_2, \ldots$ belong to the strict domain of attraction of a stable law with index α, then the scaled random walk

$$\frac{1}{n^{1/\alpha}\ell(n)}\sum_{i=1}^n X_i$$

converges weakly to a strictly α-stable law, where $\ell(n)$ is a slowly varying function; i.e. a function which satisfies $\ell(\lambda n)/\ell(n) \to 1$ as $n \to \infty$ for each fixed $\lambda > 0$. If $\ell(n)$ is a constant function, we say that $X_1, X_2, \ldots$ belong to the *strict domain of normal attraction of a stable law with index* α. Note that the classical central limit theorem states that i.i.d. random variables with mean 0 and finite variance belong to the strict domain of normal attraction of a normal law, which is a stable law with index $\alpha = 2$.

The notion of stability of random variables can be generalized to that of random vectors in $\mathbb{R}^n$ by replacing $X, X_1, \ldots, X_n$ and d_n in the above definition by vectors. Particularly interesting stable random vectors include those that are spherically symmetric. Recall that a random vector X in $\mathbb{R}^n$ is called *symmetric* if $-X \stackrel{d}{=} X$; it is called *spherically symmetric* if $UX \stackrel{d}{=} X$ for every $n \times n$-orthogonal matrix U. Clearly, the two notions agree when $n = 1$. By [Sato (1999), Thm. 14.14], the characteristic function of a spherically symmetric stable random vector X with stability index $\alpha \in (0, 2]$ takes the form

$$\mathbb{E}[e^{i\langle \xi, X \rangle}] = e^{-c|\xi|^\alpha}$$

with $c > 0$.

A Lévy process $L = (L_t)_{t \geq 0}$ in $\mathbb{R}^n$ is said to be a *stable Lévy process* with index $\alpha \in (0, 2]$ if each random vector L_t is stable with the same index α. If each L_t is strictly α-stable, then L is *self-similar* with Hurst index $1/\alpha$; i.e. for all $a > 0$,

$$(L_{at})_{t \geq 0} \stackrel{d}{=} (a^{1/\alpha} L_t)_{t \geq 0}$$

with equality in the sense of finite-dimensional distributions (see e.g. [Embrechts and Maejima (2002)]). If L is a spherically symmetric α-stable Lévy process (so that each random vector L_t is α-stable and spherically symmetric), then the characteristic function takes the form

$$\mathbb{E}[e^{i\langle \xi, L_t \rangle}] = e^{-tc|\xi|^\alpha} \tag{5.23}$$

with $c > 0$. In other words, the Lévy symbol in (5.14) is given by $\Psi(\xi) = -c|\xi|^\alpha$. In the case when $\alpha \in (0, 2)$ and $c = 1$, the backward FPK equation (5.20) becomes

$$\frac{\partial}{\partial t} u(t, x) = -\kappa_\alpha (-\Delta)^{\alpha/2} u(t, x), \ t > 0, \ x \in \mathbb{R}^n, \tag{5.24}$$

where κ_α is the diffusion constant and $-(-\Delta)^{\alpha/2}$ is interpreted as a fractional power of the Laplacian Δ. Setting $\alpha = 2$ and $c = 1/2$ recovers the FPK equation for the Brownian motion obtained in Section 2.4:

$$\frac{\partial}{\partial t} u(t, x) = \frac{1}{2} \Delta u(t, x), \ t > 0, \ x \in \mathbb{R}^n. \tag{5.25}$$

5.4   Subordinators and their inverses

One-dimensional nondecreasing Lévy processes starting at 0 are called *subordinators*. This implies Σ is the zero matrix, $b_0 \geq 0$, and $\nu(-\infty, 0) = 0$ in (5.14).

Particularly important for discussions given in Chapters 6 and 7 is the class of stable subordinators. A *stable subordinator* of index $\beta \in (0,1)$ is a one-dimensional strictly increasing stable Lévy process $U^\beta = (U_t^\beta)_{t\geq 0}$ with stability index β which is characterized by the Laplace transform

$$\mathbb{E}[e^{-sU_t^\beta}] = e^{-ts^\beta}, \quad s \geq 0. \tag{5.26}$$

The stable subordinator U^β is self-similar with Hurst index $1/\beta$. It follows from the general theory of Laplace transforms (see, e.g. [Widder (1941)]) that the density $f_{U_1^\beta}(\tau)$ of U_1^β is infinitely differentiable on $(0, \infty)$, with the following asymptotic behavior at zero and infinity (see [Mainardi et al. (2001), Uchaykin and Zolotarev (1999)]):

$$f_{U_1^\beta}(\tau) \sim \frac{\left(\frac{\beta}{\tau}\right)^{\frac{2-\beta}{2(1-\beta)}}}{\sqrt{2\pi\beta(1-\beta)}} \, e^{-(1-\beta)\left(\frac{\tau}{\beta}\right)^{-\frac{\beta}{1-\beta}}} \quad \text{as } \tau \to 0; \tag{5.27}$$

$$f_{U_1^\beta}(\tau) \sim \frac{\beta}{\Gamma(1-\beta)\tau^{1+\beta}} \quad \text{as } \tau \to \infty. \tag{5.28}$$

In particular, it follows from (5.28) that U_1^β has infinite mean; i.e. $\mathbb{E}[U_1^\beta] = \infty$. The self-similarity with the Hurst index $1/\beta$ implies that the densities of U_t^β and U_1^β are connected through the relation

$$f_{U_t^\beta}(\tau) = \frac{1}{t^{1/\beta}} f_{U_1^\beta}\left(\frac{\tau}{t^{1/\beta}}\right), \quad \tau > 0, \ t > 0. \tag{5.29}$$

The *(generalized) inverse* or the *first hitting time process* of a càdlàg, nondecreasing process $U = (U_t)_{t\geq 0}$ is a process $E = (E_t)_{t\geq 0}$ defined by

$$E_t := \inf\{\tau > 0; U_\tau > t\}.$$

It is easy to see that E is also càdlàg and nondecreasing. An *inverse stable subordinator* $E^\beta = (E_t^\beta)_{t\geq 0}$ of index $\beta \in (0,1)$ is defined to be the inverse of a stable subordinator U^β of index β. Since U^β is strictly increasing, its inverse E^β is continuous and nondecreasing. We will later observe that if U^β is adapted to a filtration $(\mathcal{F}_t)$, then E^β is a continuous $(\mathcal{F}_t)$-time-change, which is important for the purpose of discussions of stochastic integrals and SDEs driven by time-changed semimartingales (see Sections 6.1 and 6.2).

Below we discuss various properties of the inverse stable subordinator E^β of index $\beta \in (0,1)$, which can be found in [Meerschaert and Scheffler (2004)]. First, increments of E^β are neither independent nor stationary (see [Meerschaert and Scheffler (2004), Section 3]), and therefore, it is no longer a Lévy process. The inverse stable subordinator E^β is self-similar with Hurst index β. Indeed, since U^β

is self-similar with Hurst index $1/\beta$, for all $a > 0$,

$$
\begin{aligned}
\mathbb{P}(E_{at_1}^{\beta} \le \tau_1, \cdots, E_{at_n}^{\beta} \le \tau_n) &= \mathbb{P}(U_{\tau_1}^{\beta} \ge at_1, \cdots, U_{\tau_n}^{\beta} \ge at_n) \\
&= \mathbb{P}\left(\frac{1}{a}U_{\tau_1}^{\beta} \ge t_1, \cdots, \frac{1}{a}U_{\tau_n}^{\beta} \ge t_n\right) \\
&= \mathbb{P}\left(U_{\frac{\tau_1}{a^{\beta}}}^{\beta} \ge t_1, \cdots, U_{\frac{\tau_n}{a^{\beta}}}^{\beta} \ge t_n\right) \\
&= \mathbb{P}\left(E_{t_1}^{\beta} \le \frac{\tau_1}{a^{\beta}}, \cdots, E_{t_n}^{\beta} \le \frac{\tau_n}{a^{\beta}}\right) \\
&= \mathbb{P}(a^{\beta}E_{t_1}^{\beta} \le \tau_1, \cdots, a^{\beta}E_{t_n}^{\beta} \le \tau_n).
\end{aligned}
$$

The self-similarity with the Hurst index β implies that the densities of E_t^{β} and E_1^{β} are connected through the relation

$$
f_{E_t^{\beta}}(\tau) = \frac{1}{t^{\beta}} f_{E_1^{\beta}}\left(\frac{\tau}{t^{\beta}}\right), \qquad \tau > 0,\ t > 0. \tag{5.30}
$$

Moreover, the following representation holds:

$$
\begin{aligned}
f_{E_t^{\beta}}(\tau) &= \frac{\partial}{\partial \tau} P(E_t^{\beta} \le \tau) = \frac{\partial}{\partial \tau}(1 - P(U_{\tau}^{\beta} < t)) \\
&= -\frac{\partial}{\partial \tau} P\left(U_1^{\beta} < \frac{t}{\tau^{1/\beta}}\right) = -\frac{\partial}{\partial \tau}[Jf_{U_1^{\beta}}]\left(\frac{t}{\tau^{1/\beta}}\right).
\end{aligned} \tag{5.31}
$$

Performing the differentiation in the latter yields

$$
f_{E_t^{\beta}}(\tau) = -\frac{\partial}{\partial \tau} \int_0^{\frac{t}{\tau^{1/\beta}}} f_{U_1^{\beta}}(u)\,du = \frac{t}{\beta \tau^{1+1/\beta}} f_{U_1^{\beta}}\left(\frac{t}{\tau^{1/\beta}}\right). \tag{5.32}
$$

Summarizing, we have the following properties of the process E^{β}:

Proposition 5.5. *Let E^{β} be the inverse of a stable subordinator U^{β} of index $\beta \in (0,1)$.*

(1) *Sample paths of E^{β} are continuous and nondecreasing. They are not absolutely continuous with respect to Lebesgue measure.*
(2) *E^{β} is self-similar with Hurst index β.*
(3) *Increments of E^{β} are neither independent nor stationary.*
(4) *E_t^{β} has a C^{∞} density expressed as $f_{E_t^{\beta}}(\tau) = \dfrac{t}{\beta \tau^{1+1/\beta}} f_{U_1^{\beta}}\left(\dfrac{t}{\tau^{1/\beta}}\right)$.*

Unlike U_t^{β}, the random variable E_t^{β} has moments of all orders.

Proposition 5.6. *For $\nu > 0$,*

$$
\mathbb{E}[(E_t^{\beta})^{\nu}] = \frac{\Gamma(\nu + 1)}{\Gamma(\nu\beta + 1)} t^{\beta\nu}, \tag{5.33}
$$

where $\Gamma(\cdot)$ is Euler's gamma function.

Proof. First, notice that self-similarity of E^β with Hurst index β implies $\mathbb{E}[(E_t^\beta)^\nu] = \mathbb{E}[(t^\beta E_1^\beta)^\nu] = \mathbb{E}[(E_1^\beta)^\nu]t^{\beta\nu}$. Hence, it suffices to show

$$\mathbb{E}[(E_1^\beta)^\nu] = \frac{\Gamma(\nu+1)}{\Gamma(\nu\beta+1)}. \tag{5.34}$$

It follows from (5.32) that

$$f_{E_1^\beta}(u) = \frac{1}{\beta u^{1+1/\beta}} f_{U_1^\beta}\left(\frac{1}{u^{1/\beta}}\right).$$

By the change of variable $u = \tau/t^\beta$ and relation (5.29),

$$f_{E_1^\beta}\left(\frac{\tau}{t^\beta}\right) = \frac{t^{1+\beta}}{\beta\tau^{1+\frac{1}{\beta}}} f_{U_1^\beta}\left(\frac{t}{\tau^{1/\beta}}\right) = \frac{t^{1+\beta}}{\beta\tau} f_{U_\tau^\beta}(t),$$

or

$$f_{U_\tau^\beta}(t) = \beta\frac{\tau}{t^{1+\beta}} f_{E_1^\beta}\left(\frac{\tau}{t^\beta}\right). \tag{5.35}$$

Due to (5.26), it follows from equation (5.35) that

$$e^{-\tau s^\beta} = \int_0^\infty e^{-st} f_{U_\tau^\beta}(t)dt = \beta\tau \int_0^\infty \frac{e^{-st}}{t^{1+\beta}} f_{E_1^\beta}\left(\frac{\tau}{t^\beta}\right) dt.$$

Multiplying both sides by $\tau^{\nu-1}$ and then integrating over the interval $(0,\infty)$ yields

$$\int_0^\infty \tau^{\nu-1} e^{-\tau s^\beta} d\tau = \beta \int_0^\infty \tau^\nu \left[\int_0^\infty \frac{e^{-st}}{t^{1+\beta}} f_{E_1^\beta}\left(\frac{\tau}{t^\beta}\right) dt\right] d\tau. \tag{5.36}$$

For the integral on the left,

$$\int_0^\infty \tau^{\nu-1} e^{-\tau s^\beta} d\tau = \frac{\Gamma(\nu)}{s^{\beta\nu}}, \tag{5.37}$$

due to a classic formula on the Laplace transform: $L[\tau^{\nu-1}](s) = \Gamma(\nu)/s^\nu$. To compute the integral on the right, we first change the order of integration and then use the substitution $\tau = vt^\beta$ in the inner integral. Thus

$$\beta \int_0^\infty \tau^\nu \left[\int_0^\infty \frac{e^{-st}}{t^{1+\beta}} f_{E_1^\beta}\left(\frac{\tau}{t^\beta}\right) dt\right] d\tau = \beta \int_0^\infty \frac{e^{-st}}{t^{1+\beta}} \left(\int_0^\infty \tau^\nu f_{E_1^\beta}\left(\frac{\tau}{t^\beta}\right) d\tau\right) dt$$

$$= \beta \int_0^\infty t^{\beta\nu-1} e^{-st} dt \int_0^\infty v^\nu f_{E_1^\beta}(v) dv$$

$$= \frac{\beta\Gamma(\beta\nu)}{s^{\beta\nu}} \mathbb{E}[(E_1^\beta)^\nu] \tag{5.38}$$

Equating the right hand sides of (5.37) and (5.38) and using the relation $z\Gamma(z) = \Gamma(z+1)$ yields the desired result (5.34). $\qquad\square$

Below we prove a useful estimate for the density of the generalized inverse of a mixture of stable subordinators. This estimate will be used in Chapter 7. Let $U_1 = (U_{1,t})_{t\geq0}$ and $U_2 = (U_{2,t})_{t\geq0}$ be independent strictly increasing, càdlàg processes. Then the process $U = (U_t)_{t\geq0}$ defined by $U_t := U_{1,t} + U_{2,t}$ is also strictly increasing and càdlàg, and hence, the inverse E of U is continuous and nondecreasing. Moreover,

$$\mathbb{P}(E_t \leqslant \tau) = \mathbb{P}(U_\tau > t) = 1 - (F_\tau^{(1)} * F_\tau^{(2)})(t),$$

where for $k = 1, 2$, $F_\tau^{(k)}(t) = \mathbb{P}(U_{k,\tau} \leqslant t)$ with density $f_\tau^{(k)}$, and $*$ denotes convolution of cumulative distribution functions or densities, whichever is required. For notational convenience, for $a, b > 0$, let

$$\left[F_1^{(1)} \left(\frac{\cdot}{a} \right) * F_1^{(2)} \left(\frac{\cdot}{b} \right) \right](t) := \int_{s=0}^{s=t} F_1^{(1)} \left(\frac{t-s}{a} \right) dF_1^{(2)} \left(\frac{s}{b} \right),$$

which through the density functions can also be written as

$$\left[F_1^{(1)} \left(\frac{\cdot}{a} \right) * F_1^{(2)} \left(\frac{\cdot}{b} \right) \right](t) = \frac{1}{b} \int_{s=0}^{s=t} (Jf_1^{(1)}) \left(\frac{t-s}{a} \right) f_1^{(2)} \left(\frac{s}{b} \right) ds,$$

where J is the usual integration operator; i.e. $(Jg)(t) = \int_0^t g(s) ds$.

Two lemmas below are proved in [Umarov (2015b)]. Due to their importance in our discussions in Chapter 7 and for the sake of completeness, we reproduce their proofs.

Lemma 5.3. *Let U_1 and U_2 be independent stable subordinators with respective indices β_1 and β_2 in $(0,1)$. Let c_1 and c_2 be positive constants. Define a process U by $U_t := c_1 U_{1,t} + c_2 U_{2,t}$ for $t \geqslant 0$. Then the inverse E of U satisfies*

$$\mathbb{P}(E_t \leqslant \tau) = 1 - \left[F_1^{(1)} \left(\frac{\cdot}{c_1 \tau^{\frac{1}{\beta_1}}} \right) * F_1^{(2)} \left(\frac{\cdot}{c_2 \tau^{\frac{1}{\beta_2}}} \right) \right](t) \qquad (5.39)$$

and has density

$$f_{E_t}(\tau) = -\frac{\partial}{\partial \tau} \left\{ \frac{1}{c_2 \tau^{\frac{1}{\beta_2}}} \left[(Jf_1^{(1)}) \left(\frac{\cdot}{c_1 \tau^{\frac{1}{\beta_1}}} \right) * f_1^{(2)} \left(\frac{\cdot}{c_2 \tau^{\frac{1}{\beta_2}}} \right) \right](t) \right\}. \qquad (5.40)$$

Proof. Since U_1 and U_2 are independent and self-similar with respective Hurst indices $1/\beta_1$ and $1/\beta_2$,

$$\mathbb{P}(E_t \leqslant \tau) = \mathbb{P}(U_\tau > t) = 1 - \mathbb{P}\left(c_1 \tau^{\frac{1}{\beta_1}} U_{1,1} + c_2 \tau^{\frac{1}{\beta_2}} U_{2,1} \leqslant t \right)$$

$$= 1 - \left[F_1^{(1)} \left(\frac{\cdot}{c_1 \tau^{\frac{1}{\beta_1}}} \right) * F_1^{(2)} \left(\frac{\cdot}{c_2 \tau^{\frac{1}{\beta_2}}} \right) \right](t),$$

from which (5.40) follows immediately upon differentiating with respect to τ.   □

Lemma 5.4. *Let U_k, $k = 1, \ldots, N$, be independent stable subordinators with respective indices $\beta_k \in (0,1)$, $k = 1, \ldots, N$. Let c_k, $k = 1, \ldots, N$, be positive constants. Define a process U by $U_t := \sum_{k=1}^{N} c_k U_{k,t}$ for $t \geqslant 0$. Let E be the inverse of U. Then for any $t \in (0, \infty)$, the density $f_{E_t}(\tau)$ of E_t is bounded and there exist a number $\beta \in (0,1)$ and positive constants C, K, not depending on τ, such that*

$$f_{E_t}(\tau) \leqslant C \exp\left(-K \tau^{\frac{1}{1-\beta}} \right) \qquad (5.41)$$

for τ large enough.

Proof. It suffices to prove the lemma for $N = 2$. The general case then follows by induction. Suppose for clarity that $0 < \beta_1 < \beta_2 < 1$ in representation (5.40). It follows that $f_{E_t}(\tau) = I_1 + I_2 + I_3$, where

$$I_1 = \frac{1}{\beta_2 c_2 \tau^{1+\frac{1}{\beta_2}}} \int_0^t (Jf_1^{(1)}) \left(\frac{s}{c_1 \tau^{\frac{1}{\beta_1}}}\right) f_1^{(2)} \left(\frac{t-s}{c_2 \tau^{\frac{1}{\beta_2}}}\right) ds, \tag{5.42}$$

$$I_2 = \frac{1}{\beta_1 c_1 c_2 \tau^{1+\frac{1}{\beta_1}+\frac{1}{\beta_2}}} \int_0^t s \cdot f_1^{(1)} \left(\frac{s}{c_1 \tau^{\frac{1}{\beta_1}}}\right) f_1^{(2)} \left(\frac{t-s}{c_2 \tau^{\frac{1}{\beta_2}}}\right) ds, \tag{5.43}$$

and

$$I_3 = \frac{1}{c_2 \tau^{\frac{1}{\beta_2}}} \int_0^t s \cdot (Jf_1^{(1)}) \left(\frac{s}{c_1 \tau^{\frac{1}{\beta_1}}}\right) (f_1^{(2)})' \left(\frac{t-s}{c_2 \tau^{\frac{1}{\beta_2}}}\right) ds. \tag{5.44}$$

It is easy to see that integration by parts reduces I_3 to the sum of integrals of types I_1 and I_2, namely, $I_3 = \beta_2 c_2 \tau^{1+\frac{1}{\beta_2}} I_1 + \beta_1 \tau I_2$. Therefore, it suffices to estimate I_1 and I_2. First, notice that both functions $f_1^{(1)}$, $f_1^{(2)}$ are continuous on $[0, \infty)$, and $Jf_1^{(1)}(t) \leqslant 1$. Consequently, in accordance with the mean value theorem, there exist numbers $s_*, s_{**} \in (0, t)$ such that

$$I_1 \leqslant \frac{t}{\beta_2 c_2 \tau^{1+\frac{1}{\beta_2}}} f_1^{(2)} \left(\frac{s_*}{c_2 \tau^{\frac{1}{\beta_2}}}\right), \tag{5.45}$$

and

$$I_2 = \frac{t s_{**}}{\beta_1 c_1 c_2 \tau^{1+\frac{1}{\beta_1}+\frac{1}{\beta_2}}} f_1^{(1)} \left(\frac{s_{**}}{c_1 \tau^{\frac{1}{\beta_1}}}\right) f_1^{(2)} \left(\frac{t-s_{**}}{c_2 \tau^{\frac{1}{\beta_2}}}\right). \tag{5.46}$$

For τ small enough, (5.28) implies

$$I_1 \leqslant C_1, \quad I_2 \leqslant C_2 \tau \quad \text{and} \quad I_3 \leqslant C_3 \tau^2,$$

where C_1, C_2 and C_3 are constants not depending on τ. These estimates and continuity of convolution imply boundedness of $f_{E_t}(\tau)$ for any $\tau < \infty$.

Now suppose that τ is large enough. Then taking into account (5.27) in (5.45) and (5.46), it is not hard to verify that

$$I_1 \leqslant \frac{C_3}{\tau^{\frac{1-2\beta_2}{2(1-\beta_2)}}} \exp\left(-K_1 \tau^{\frac{1}{1-\beta_2}}\right) \tag{5.47}$$

and

$$I_2 \leqslant \frac{C_4}{\tau^{1-\frac{\beta_1}{2(1-\beta_1)}-\frac{\beta_2}{2(1-\beta_2)}}} \exp\left(-K_2(\tau^{\frac{1}{1-\beta_1}} + \tau^{\frac{1}{1-\beta_2}})\right), \tag{5.48}$$

where C_3, C_4, K_1 and K_2 are positive constants not depending on τ. Letting $\beta = \beta_1 = \min(\beta_1, \beta_2)$, $C = \max(C_3, C_4)$, and $K = \min(K_1, 2K_2) - \varepsilon$, where $\varepsilon \in (0, \min(K_1, 2K_2))$, yields (5.41). $\qquad\square$

Remark 5.1. (a) Further properties of E_t^β and its density function $f_{E_t^\beta}(\tau)$ will be studied in Section 7.3.

(b) Tempered stable subordinators, as well as stable subordinators, have found many important applications recently. In the simplest case, the distribution of a tempered stable subordinator $U^{\beta,\lambda}$ involves a tempering parameter $\lambda > 0$ in addition to the stability index $\beta \in (0,1)$ of an underlying stable subordinator U^β. The density of $U^{\beta,\lambda}$ is written in terms of that of U^β as

$$f_{U_t^{\beta,\lambda}}(\tau) = e^{-\lambda\tau + \lambda^\beta t} f_{U_t^\beta}(\tau), \quad \tau \geq 0$$

and the Laplace transform is given by

$$\mathbb{E}[e^{-sU_t^{\beta,\lambda}}] = e^{-t[(s+\lambda)^\beta - \lambda^\beta]}.$$

Inverses of tempered stable subordinators are studied in [Kumar and Vellaisamy (2015)].

(c) A thorough discussion of a very general class of multivariate tempered stable Lévy processes is given in [Rosiński (2007)], where tempered stable distributions are obtained by tempering stable Lévy measures as follows. A *tempered α-stable distribution* on $\mathbb{R}^d$ is defined to be an infinitely divisible distribution without Gaussian part and with Lévy measure ν written in polar coordinates as

$$\nu(dr, du) = \frac{q(r, u)}{r^{1+\alpha}}\, dr\, \sigma(du),$$

where $\alpha \in (0,2)$, σ is a finite Borel measure on S^{d-1}, and $q : (0,\infty) \times S^{d-1} \to (0,\infty)$ is a Borel measurable function such that for each fixed $u \in S^{d-1}$, $q(\cdot, u)$ is a completely monotone function vanishing at ∞. A tempered stable Lévy process is a Lévy process $(X_t)_{t\geq 0}$ with X_1 having a tempered stable distribution. In [Rosiński (2007), Thm. 3.1], it is shown that a tempered stable process looks like a stable process in a short time scale and a Brownian motion in a large time scale. The paper also investigates a deep connection of the distributions of tempered stable processes with those of the underlying stable processes. It is also worth noting that [Rosiński (2007), Prop. 2.7(iv)] particularly implies that the tempered stable subordinator $U^{\beta,\lambda}$ discussed above has moments of all orders, unlike the stable subordinator U^β.

5.5   Gaussian processes

This short section introduces the basic notion of Gaussian processes and some of their examples that are important in various applications. Fokker–Planck–Kolmogorov equations associated with these processes and their time-changed versions will be derived in Sections 7.3–7.7.

A one-dimensional stochastic process $X = (X_t)_{t\geq 0}$ is called a *Gaussian process* if the random vector $(X_{t_1}, \ldots, X_{t_m})$ has a multivariate Gaussian distribution for all finite sequences $0 \leq t_1 < \cdots < t_m < \infty$. The joint distributions are characterized by the mean function $\mathbb{E}[X_t]$ and the covariance function $R_X(s,t) = \text{Cov}(X_s, X_t)$. In this book, for simplicity, we consider zero-mean Gaussian processes; i.e. those for which $\mathbb{E}[X_t] = 0$ for all $t \geqslant 0$. The class of Gaussian processes contains some of the most important stochastic processes in both theoretical and applied probability, including Brownian motion, fractional Brownian motion [Biagini et al. (2008), Cheridito (2003), Fannjiang and Komorowski (2000), Silbergleit et al. (2007)], and Volterra processes [Alòs et al. (2001), Decreusefond (2005)].

A one-dimensional *fractional Brownian motion (fBM)* $B^H = (B_t^H)_{t \geqslant 0}$ is a zero-mean Gaussian process with continuous paths and covariance function

$$R_{B^H}(s,t) = \mathbb{E}[B_s^H B_t^H] = \frac{1}{2}(s^{2H} + t^{2H} - |s-t|^{2H}), \qquad (5.49)$$

where $H \in (0,1)$ is called the Hurst parameter (as B^H is self-similar with Hurst index H). If $H = 1/2$, then B^H becomes a usual Brownian motion since it follows that $R_{B^H}(s,t) = \min(s,t)$. Fractional Brownian motion B^H has many properties similar to those of Brownian motion. For example, it has stationary increments and nowhere differentiable but Hölder continuous sample paths of any order less than H. However, it does not have independent increments. Namely, the covariance between increments over non-overlapping intervals is positive if $1/2 < H < 1$, and negative if $0 < H < 1/2$. In particular, when $1/2 < H < 1$, the increments exhibit long range dependence and B^H has the integral representation

$$B_t^H = \int_0^t K_H(t,s) dW_s,$$

where W is a Brownian motion and $K_H(t,s)$ is a deterministic kernel given by

$$K_H(t,s) = c_H s^{1/2-H} \int_s^t (r-s)^{H-3/2} r^{H-1/2} dr, \quad t > s.$$

Here, the positive constant c_H is chosen so that the integral $\int_0^{t \wedge s} K_H(t,r) K_H(s,r) dr$ coincides with $R_{B^H}(s,t)$ in (5.49). Fractional Brownian motion is not a semimartingale unless $H = 1/2$ (see [Biagini et al. (2008), Nualart (2006)]), so the usual Itô stochastic calculus discussed in Section 5.2 is not valid. Nevertheless, there are several approaches [Bender (2003), Biagini et al. (2008), Decreusefond and Uştünel (1998), Nualart (2006)] to defining stochastic integrals driven by fBM. Stochastic processes driven by fBM are of increasing interest for both theorists and applied researchers due to their wide application in fields such as mathematical finance [Cheridito (2003), Shiryaev (1999)], solar activities [Silbergleit et al. (2007)], turbulence [Fannjiang and Komorowski (2000)], etc.

Volterra processes are obtained by generalizing the definition of fBM. In short, they are continuous zero-mean Gaussian processes $V = (V_t)_{t \in [0,T]}$ with integral representations of the form

$$V_t = \int_0^t K(t,s) \, dW_s, \qquad (5.50)$$

where $K(t,s)$ is a square integrable kernel and W is Brownian motion. In [De-creusefond (2005)], a Volterra process is constructed as follows. Suppose that K_0 is a Hilbert–Schmidt linear operator from $L^2[0,1]$ into itself satisfying some hypothe-ses including the triangularity of K_0. There exists a measurable kernel $K(t,s)$ with $K(t,s) = 0$ for $s > t$ such that

$$K_0 f(t) = \int_0^t K(t,s)f(s)\,ds.$$

The kernel $R(t,s)$ defined by

$$R(t,s) = \int_0^{t\wedge s} K(t,r)K(s,r)\,dr$$

is shown to be nonnegative definite, and therefore, we can discuss a zero-mean Gaus-sian process V with covariance function $R(t,s)$. The process V has a continuous modification and has an integral representation of the form (5.50); see [Decreusefond (2005)] for details. A different construction of Volterra processes is given in [Alòs et al. (2001)], where assumptions are placed on the kernel $K(t,s)$ rather than on the operator K_0. Both papers [Alòs et al. (2001), Decreusefond (2005)] define stochastic integrals driven by Volterra processes using Malliavin calculus and establish perti-nent Itô formulae. The simplest version of the Itô formula provided in [Alòs et al. (2001)] takes the form

$$f(V_t) = f(0) + \int_0^t f'(V_s)\,\delta V_s + \frac{1}{2}\int_0^t f''(V_s)\,dR_s,$$

where $R_s := R(s,s) = \mathbb{E}[V_s^2]$ is the variance function of V and $\int_0^t f'(V_s)\,\delta V_s$ is a stochastic integral defined via the divergence operator δ with respect to V. For more details about Volterra processes and associated stochastic integration, consult [Alòs et al. (2001), Decreusefond (2005)].

Chapter 6

Stochastic calculus for time-changed semimartingales and its applications to SDEs

Introduction

Section 5.2 summarized important properties of stochastic integrals driven by semi-martingales, followed by a discussion of time-changed processes. Namely, time-changed semimartingales are again semimartingales, but with respect to a time-changed filtration. On the other hand, time-changed local martingales are local martingales provided the local martingales are in synchronization with the time-change. This chapter begins by investigating

(1) the notion of synchronization in connection with stochastic integrals driven by time-changed semimartingales, and

(2) stochastic differential equations (SDEs) that those time-changed processes drive.

The first two sections provide important theoretical tools for proving some of the main results of this book in Chapter 7. The next three sections focus on approximations using continuous time random walks (CTRWs). Section 6.3 discusses how driving processes which are Lévy processes time-changed by an inverse stable subordinator can arise as scaling limits of CTRWs. An example given at the end of this section motivates the subject of Section 6.4, namely connections of CTRW approximations to fractional order differential equations. Section 6.3 is also connected to the subject of Section 6.5, namely approximation of stochastic integrals driven by scaling limits of CTRWs. The final Section 6.6 focuses on numerical approximation of SDEs driven by a time-changed Brownian motion. These approximation methods disclose the significance of "CTRW" in our original diagram presented in Figure 1.1 in Chapter 1.

6.1 Stochastic calculus for time-changed semimartingales

Recall that a process $Z = (Z_t)$ is said to be *in synchronization* with a time-change $E = (E_t)$ if Z is constant on every interval of the form $[E_{t-}, E_t]$ a.s. We occasionally write $Z \sim_{\text{synch}} E$ for shorthand. Recall also that $L(Z_t, \mathcal{F}_t)$ denotes the class of $(\mathcal{F}_t)$-predictable processes H for which the Itô-type stochastic integral $H \bullet Z$ is defined. One quite simple yet significant observation, which connects the notion of synchronization with stochastic integrals, is that if an $(\mathcal{F}_t)$-semimartingale Z is in synchronization with a finite $(\mathcal{F}_t)$-time-change E and if $H \in L(Z_t, \mathcal{F}_t)$, then $(H_{E(t-)}) \in L(Z_{E_t}, \mathcal{F}_{E_t})$, where $H_{E(t-)}$ denotes the process H evaluated at the left limit point E_{t-} of E at t. This observation leads to the consideration of two integral processes $(\int_0^t H_s dZ_s$ and $\int_0^t H_{E(s-)} dZ_{E_s})$. Since stochastic integrals of semimartingales are again semimartingales adapted to the same filtration, the two integrals $(\int_0^t H_s dZ_s)$ and $(\int_0^t H_{E(s-)} dZ_{E_s})$ are semimartingales with respect to the filtrations $(\mathcal{F}_t)$ and $(\mathcal{F}_{E_t})$, respectively. By Lemma 5.2, the former stochastic integral can be time-changed by E to produce another $(\mathcal{F}_{E_t})$-semimartingale. The two $(\mathcal{F}_{E_t})$-semimartingales $(\int_0^{E_t} H_s dZ_s)$ and $(\int_0^t H_{E(s-)} dZ_{E_s})$ coincide for any $H \in L(Z_t, \mathcal{F}_t)$. This fact plays a significant role in establishing Lemma 6.5; hence, it is stated here as a lemma.

Lemma 6.1. *(1st Change-of-Variable Formula)* [Jacod (1979), Prop. 10.21] *Let Z be an $(\mathcal{F}_t)$-semimartingale which is in synchronization with a finite $(\mathcal{F}_t)$-time-change E. If $H \in L(Z_t, \mathcal{F}_t)$, then $(H_{E(t-)}) \in L(Z_{E_t}, \mathcal{F}_{E_t})$. Moreover, with probability one, for all $t \geqslant 0$,*

$$\int_0^{E_t} H_s dZ_s = \int_0^t H_{E(s-)} dZ_{E_s}. \tag{6.1}$$

Recall that the unique continuous local martingale part of a semimartingale Z is denoted Z^c (see Section 5.2).

Lemma 6.2. [Jacod (1979), Thm. 10.17] *Let Z be a one-dimensional $(\mathcal{F}_t)$-semimartingale which is in synchronization with a finite $(\mathcal{F}_t)$-time-change E. Then Z^c and $[Z, Z]$ are also in synchronization with E. Moreover,*

$$[Z \circ E, Z \circ E] = [Z, Z] \circ E; \tag{6.2}$$
$$(Z \circ E)^c = Z^c \circ E. \tag{6.3}$$

The following simple example explains the significance of the synchronization assumption in Lemmas 6.1 and 6.2. Note that since Brownian sample paths never stay flat on any time interval, Brownian motion cannot be in synchronization with any time-change having discontinuous paths.

Example 6.1. Let $Z = B$ be an $(\mathcal{F}_t)$-Brownian motion, and define a deterministic time-change E by $E_t := \boldsymbol{I}_{[1,\infty)}(t)$, where $\boldsymbol{I}_A$ denotes the indicator function of a set A. Clearly, B is not in synchronization with E. Let H be a deterministic process given by $H_t = \boldsymbol{I}_{(1/2,\,\infty)}(t)$, then $H_{E(t-)} = \boldsymbol{I}_{(1,\infty)}(t)$. Hence,

$$\int_0^{E_1} H_s dB_s = \int_0^1 H_s dB_s = \int_{1/2}^1 dB_s = B_1 - B_{1/2};$$
$$\int_0^1 H_{E(s-)} dB_{E_s} = \int_0^1 0\, dB_{E_s} = 0.$$

Therefore, the two integrals in (6.1) fail to coincide. Moreover, it follows from (5.10) that

$$[B \circ E, B \circ E]_1 = (B_{E_1})^2 - 2 \int_0^1 B_{E_{s-}} dB_{E_s} = B_1^2 - 2 \int_0^1 0 \, dB_{E_s} = B_1^2,$$

whereas $([B, B] \circ E)_1 = E_1 = 1$. Therefore, (6.2) does not hold. Furthermore, since $B \circ E$ is not a continuous process, $(B \circ E)^c$ and $B^c \circ E \ (= B \circ E)$ fail to coincide. Thus, (6.3) does not hold either.

The next lemma will be used in the proof of Lemma 6.5.

Lemma 6.3. *Let Z be an $(\mathcal{F}_t)$-semimartingale which is in synchronization with a finite $(\mathcal{F}_t)$-time-change E. Let $H \in L(Z_t, \mathcal{F}_t)$. Then the stochastic integral $H \bullet Z$ is also in synchronization with E.*

Proof. Fix $t \geqslant 0$, and let $u \in [E_{t-}, E_t]$. Since $Z \sim_{\text{synch}} E$, Z is constant on $[u, E_t]$; hence, $(H \bullet Z)_{E_t} - (H \bullet Z)_u = \int_{u+}^{E_t} H_s dZ_s = 0$. Therefore, $(H \bullet Z)_{E_t} = (H \bullet Z)_u$. Thus, $H \bullet Z$ is constant on $[E_{t-}, E_t]$. □

Recall that the *(generalized) inverse* of a càdlàg, nondecreasing process U is a process E defined by

$$E_t := \inf\{\tau > 0; U_\tau > t\}.$$

It is easy to see that E is also càdlàg and nondecreasing. Note that every $(\mathcal{F}_t)$-adapted, càdlàg, nondecreasing process has paths of finite variation on compact sets; hence, *a priori* it is an $(\mathcal{F}_t)$-semimartingale.

Lemma 6.4. *(1) Let U be a nondecreasing $(\mathcal{F}_t)$-semimartingale such that $\lim_{t\to\infty} U_t = \infty$. Then the inverse E of U is a finite $(\mathcal{F}_t)$-time-change such that $\lim_{t\to\infty} E_t = \infty$. Moreover, if U is strictly increasing, then E has continuous paths.*

(2) Let E be a finite $(\mathcal{F}_t)$-time-change such that $\lim_{t\to\infty} E_t = \infty$. Then the inverse U of E is a nondecreasing $(\mathcal{F}_t)$-semimartingale such that $\lim_{t\to\infty} U_t = \infty$. Moreover, if E has continuous paths, then U is strictly increasing.

Proof. (1) The assumption $\lim_{t\to\infty} U_t = \infty$ implies that each random variable E_t is finite. In addition, since each U_t is real-valued, it follows that $\lim_{t\to\infty} E_t = \infty$. Fix $t \geqslant 0$. Since U is $(\mathcal{F}_t)$-adapted, $\{E_t < s\} = \{U_{s-} > t\} \in \mathcal{F}_{s-} \subset \mathcal{F}_s$ for any $s > 0$, and obviously $\{E_t < 0\} = \varnothing \in \mathcal{F}_0$. Hence, it follows from the right-continuity of $(\mathcal{F}_t)$ that the random variable E_t is an $(\mathcal{F}_t)$-stopping time (see [Karatzas and Shreve (1991), Prop. 1.2.3]). Thus, E is a finite $(\mathcal{F}_t)$-time-change. Moreover, if U is strictly increasing, then E obviously has continuous paths.

(2) The assumption $\lim_{t\to\infty} E_t = \infty$ implies that each random variable U_t is finite. In addition, since each E_t is real-valued, it follows that $\lim_{t\to\infty} U_t = \infty$. Fix $s \geqslant 0$. For any $t > 0$, since E_{t-} is also an $(\mathcal{F}_t)$-stopping time, $\{U_s \geqslant t\} =$

$\{E_{t-} \leqslant s\} \in \mathcal{F}_s$. Also, $\{U_s \geqslant 0\} = \Omega \in \mathcal{F}_s$. Hence, U_s is $\mathcal{F}_s$-measurable. Therefore, U is $(\mathcal{F}_t)$-adapted. Since U is also càdlàg and nondecreasing, it is an $(\mathcal{F}_t)$-semimartingale. Moreover, if E has continuous paths, then it is clear that U is strictly increasing. $\qquad\square$

Notation 6.1. Lemma 6.4 establishes that a nondecreasing $(\mathcal{F}_t)$-semimartingale U and a finite $(\mathcal{F}_t)$-time-change E are 'dual' in the sense that either process with the specified condition induces the other. In light of this, we say a pair U and E satisfies $[U \longmapsto E]$ (or $[U \longleftarrow\!\!\!\shortmid E]$) if U is a nondecreasing $(\mathcal{F}_t)$-semimartingale and E is a finite $(\mathcal{F}_t)$-time-change for which U induces E (or E induces U). If U is strictly increasing and E has continuous paths, then the double brackets $[[U \longmapsto E]]$ and $[[U \longleftarrow\!\!\!\shortmid E]]$ are employed instead. Hence, the double bracket notation assumes stronger conditions than the single bracket notation.

The following theorem, at first glance, may seem quite simple, but its impact on the formulation of stochastic calculus for time-changed semimartingales is profound. Lemma 6.5 guarantees that *any* stochastic integral driven by a time-changed semimartingale is a time-changed stochastic integral driven by the original semimartingale, as long as the semimartingale is in synchronization with the time-change.

Lemma 6.5. *(2nd Change-of-Variable Formula) Let Z be an $(\mathcal{F}_t)$-semimartingale. Let U and E be a pair satisfying $[U \longmapsto E]$ or $[U \longleftarrow\!\!\!\shortmid E]$. Suppose Z is in synchronization with E. If $K \in L(Z_{E_t}, \mathcal{F}_{E_t})$, then $\big(K_{U(t-)}\big) \in L(Z_t, \mathcal{F}_{E(U_t)})$. Moreover, with probability one, for all $t \geqslant 0$,*

$$\int_0^t K_s dZ_{E_s} = \int_0^{E_t} K_{U(s-)} dZ_s. \tag{6.4}$$

Proof. By Lemma 5.2, both E and $X := Z \circ E$ are $(\mathcal{F}_{E_t})$-semimartingales. Since E is a nondecreasing $(\mathcal{F}_{E_t})$-semimartingale such that $\lim_{t \to \infty} E_t = \infty$ and $E_0 = 0$, it follows from Lemma 6.4 (1) that U is a finite $(\mathcal{F}_{E_t})$-time-change. On any half open interval $[U_{s-}, U_s)$, E is obviously constant by construction and hence so is X. Moreover, since $Z \sim_{\text{synch}} E$,

$$(Z \circ E)_{U(s)} = Z_{E(U(s))} = Z_{E(U(s)-)} = Z_{E(U(s-))} = (Z \circ E)_{U(s-)}.$$

Hence, $X_{U_s} = X_{U(s-)}$. Thus, X is constant on any closed interval $[U_{s-}, U_s]$. Therefore, $X \sim_{\text{synch}} U$.

Now, let $K \in L(X_t, \mathcal{F}_{E_t})$. Then it follows from Lemma 6.1 that $\big(K_{U(t-)}\big) \in L(X_{U_t}, \mathcal{F}_{E(U_t)})$. By the 1st change-of-variable formula (6.1) and the assumption $Z \sim_{\text{synch}} E$, with probability one, for all $t \geqslant 0$,

$$\int_0^{U_t} K_s dX_s = \int_0^t K_{U(s-)} dX_{U_s} = \int_0^t K_{U(s-)} dZ_{E(U(s))} = \int_0^t K_{U(s-)} dZ_s,$$

which yields

$$\int_0^{U_{E_t}} K_s dX_s = \int_0^{E_t} K_{U(s-)} dZ_s. \tag{6.5}$$

Since $X \sim_{\text{synch}} U$, Lemma 6.3 yields $K \bullet X \sim_{\text{synch}} U$. Any t is contained in the interval $[U_{E(t)-}, U_{E_t}]$, so $(K \bullet X)_{U_{E_t}} = (K \bullet X)_t$. Thus, (6.5) establishes (6.4). $\square$

If a pair U and E satisfies $[[U \longmapsto E]]$ or $[[U \longleftarrow E]]$, then *any* process Z is automatically in synchronization with E due to the continuity of E. Therefore, under either of these stronger conditions, Lemma 6.5 is valid for an arbitrary $(\mathcal{F}_t)$-semimartingale Z. Moreover, the Itô formula for stochastic integrals driven by a time-changed semimartingale can be reformulated in a nice way as in the following theorem.

Theorem 6.1. *(Time-changed Itô Formula) Let Z be a one-dimensional $(\mathcal{F}_t)$-semimartingale. Let U and E be a pair satisfying $[[U \longmapsto E]]$ or $[[U \longleftarrow E]]$. Let X be a process defined by*

$$X_t := \int_0^t A_s ds + \int_0^t F_s dE_s + \int_0^t G_s dZ_{E_s} \tag{6.6}$$

where $A \in L(t, \mathcal{F}_{E_t})$, $F \in L(E_t, \mathcal{F}_{E_t})$, and $G \in L(Z_{E_t}, \mathcal{F}_{E_t})$. If $f \in C^2(\mathbb{R})$, then $f(X)$ is an $(\mathcal{F}_{E_t})$-semimartingale, and with probability one, for all $t \geq 0$,

$$f(X_t) - f(0) = \int_0^t f'(X_{s-}) A_s ds + \int_0^{E_t} f'(X_{U(s-)-}) F_{U(s-)} ds \tag{6.7}$$
$$+ \int_0^{E_t} f'(X_{U(s-)-}) G_{U(s-)} dZ_s + \frac{1}{2} \int_0^{E_t} f''(X_{U(s-)-}) \{G_{U(s-)}\}^2 d[Z, Z]_s^c$$
$$+ \sum_{0 < s \leq t} \{f(X_s) - f(X_{s-}) - f'(X_{s-}) \Delta X_s\}.$$

In particular, if Z is a standard Brownian motion B, then with probability one, for all $t \geq 0$,

$$f(X_t) - f(0) = \int_0^t f'(X_s) A_s ds + \int_0^{E_t} f'(X_{U(s-)}) F_{U(s-)} ds \tag{6.8}$$
$$+ \int_0^{E_t} f'(X_{U(s-)}) G_{U(s-)} dB_s + \frac{1}{2} \int_0^{E_t} f''(X_{U(s-)}) \{G_{U(s-)}\}^2 ds.$$

Proof. Since the process X is defined to be a sum of stochastic integrals driven by $(\mathcal{F}_{E_t})$-semimartingales, X itself is also an $(\mathcal{F}_{E_t})$-semimartingale. The Itô formula (5.12) with $d = 1$ yields, for all $t \geq 0$,

$$f(X_t) - f(0) = \int_0^t f'(X_{s-}) dX_s + \frac{1}{2} \int_0^t f''(X_{s-}) d[X, X]_s^c \tag{6.9}$$
$$+ \sum_{0 < s \leq t} \{f(X_s) - f(X_{s-}) - f'(X_{s-}) \Delta X_s\}.$$

Using the 2nd change-of-variable formula (6.4), the first integral on the right-hand side of (6.9) becomes

$$\int_0^t f'(X_{s-})dX_s = \int_0^t f'(X_{s-})A_s ds + \int_0^t f'(X_{s-})F_s dE_s + \int_0^t f'(X_{s-})G_s dZ_{E_s}$$

(6.10)

$$= \int_0^t f'(X_{s-})A_s ds + \int_0^{E_t} f'(X_{U(s-)-})F_{U(s-)} ds + \int_0^{E_t} f'(X_{U(s-)-})G_{U(s-)} dZ_s.$$

Hence, to obtain (6.7) from (6.9), we only need to verify that

$$\int_0^t f''(X_{s-})d[X,X]_s^c = \int_0^{E_t} f''(X_{U(s-)-})\{G_{U(s-)}\}^2 d[Z,Z]_s^c.$$

(6.11)

Let $\tilde{X}_t = Z_{E_t}$ so that $X = A \bullet m + F \bullet E + G \bullet \tilde{X}$, where m is the identity map corresponding to the Lebesgue measure. We claim that

$$[X,X]_t^c = \int_0^t G_s^2 d[\tilde{X},\tilde{X}]_s^c.$$

(6.12)

To prove this, first note that m and E are both continuous processes of finite variation on compact sets. By [Protter (2004), II. Thm. 26],

$$[m,\tilde{X}]_t = \sum_{0 < s \leqslant t} \Delta[m,\tilde{X}]_s = \sum_{0 < s \leqslant t} (\Delta m_s) \cdot (\Delta \tilde{X}_s) = 0$$

for all $t \geqslant 0$. Hence, $[m,\tilde{X}] = 0$. Similarly, $[m,m] = [m,E] = [E,E] = [E,\tilde{X}] = 0$. Therefore, the bilinearity of $[\cdot,\cdot]$ and Lemma 5.1 (2) imply

$$[X,X] = [A \bullet m + F \bullet E + G \bullet \tilde{X}, A \bullet m + F \bullet E + G \bullet \tilde{X}] = G^2 \bullet [\tilde{X},\tilde{X}]. \quad (6.13)$$

Now, let $J_t := \sum_{0 < s \leqslant t} \Delta[\tilde{X},\tilde{X}]_s$ so that $[\tilde{X},\tilde{X}]_t^c = [\tilde{X},\tilde{X}]_t - J_t$. Then the pure jump process J shares with $[\tilde{X},\tilde{X}]$ the same jump times and sizes. Therefore,

$$\sum_{0 < s \leqslant t} G_s^2 \Delta[\tilde{X},\tilde{X}]_s = \sum_{0 < s \leqslant t} G_s^2 \Delta J_s = \int_0^t G_s^2 dJ_s.$$

Hence, it follows from (6.13) together with Lemma 5.1 (1) that

$$[X,X]_t^c = [X,X]_t - \sum_{0 < s \leqslant t} \Delta[X,X]_s = (G^2 \bullet [\tilde{X},\tilde{X}])_t - \sum_{0 < s \leqslant t} G_s^2 \Delta[\tilde{X},\tilde{X}]_s$$

$$= \int_0^t G_s^2 d[\tilde{X},\tilde{X}]_s - \int_0^t G_s^2 dJ_s = \int_0^t G_s^2 d[\tilde{X},\tilde{X}]_s^c,$$

thereby establishing (6.12).

Since $Z \sim_{\text{synch}} E$, repeated use of Lemma 6.2 together with identity (5.11) yields

$$[\tilde{X},\tilde{X}]^c = [\tilde{X}^c,\tilde{X}^c] = [Z^c \circ E, Z^c \circ E] = [Z^c,Z^c] \circ E = [Z,Z]^c \circ E. \quad (6.14)$$

Together (6.12) and (6.14) yield $[X,X]_t^c = \int_0^t G_s^2 d[Z,Z]_{E_s}^c$. Therefore, (6.11) follows immediately from the 2nd change-of-variable formula (6.4) and the proof of the theorem is complete. $\square$

A similar proof yields the multidimensional version of Theorem 6.1, the statement of which is found in [Kobayashi (2011)]. A version of Theorem 6.1 appears in [Nane and Ni (2017)]. In [Nane and Ni (2017), Wu (2016)], the time-changed Itô formula is used to analyze stability of solutions of stochastic differential equations driven by time-changed processes.

Remark 6.1. (a) The first integral in Formula (6.7) can also be expressed as a time-changed stochastic integral. By the 2nd change-of-variable formula (6.4),

$$\int_0^t f'(X_{s-})A_s ds = \int_0^t f'(X_{s-})A_s dU_{E_s} + \sum_{0<s\leqslant t} f'(X_{s-})A_s \Delta(U \circ E)_s \qquad (6.15)$$

$$= \int_0^{E_t} f'(X_{U(s-)-})A_{U(s-)} dU_s + \sum_{0<s\leqslant t} f'(X_{s-})A_s \Delta(U \circ E)_s$$

as long as all integrals are defined. The additional term arises due to the discontinuities of U.

(b) The stronger condition $[[U \longmapsto E]]$ or $[[U \longleftarrow E]]$, rather than $[U \longmapsto E]$ or $[U \longleftarrow E]$, is essential in establishing the nice representations (6.7) and (6.8). For example, if E has jumps, then the stochastic integral $\int_0^t f'(X_{s-})F_s dE_s$ in (6.10) may not be rephrased as a time-changed integral driven by ds since the identity map $m(s) = s$ is no longer in synchronization with E. Moreover, the equalities $[E, E] = 0$ and $[E, \tilde{X}] = 0$ both may fail, which implies more terms need to be included in (6.12).

(c) In real situations, the distributions of Z, U and E are chosen to model statistical data, and scientists will seek to reveal the behavior of a process X described via a stochastic differential equation (SDE) of the form

$$dX_t = \rho(t, E_t, X_t)dt + \mu(t, E_t, X_t)dE_t + \sigma(t, E_t, X_t)dZ_{E_t}. \qquad (6.16)$$

Formula (6.7) encourages handling the solution of (6.16) via conditioning. In particular, when Z is continuous and $A \equiv 0$, the right hand side of formula (6.7), conditioned on E_t, can be regarded as usual stochastic integrals driven simply by Lebesgue measure, Z and $[Z, Z]$. In fact, in Chapter 7, the time-changed Itô formula (6.8) together with a conditioning argument reveals a connection between the class of SDEs driven by a time-changed Brownian motion and the associated class of Fokker–Planck–Kolmogorov equations.

6.2 SDEs driven by time-changed semimartingales

A classical Itô SDE is of the form

$$dY_t = b(t, Y_t)dt + \tau(t, Y_t)dB_t \quad \text{with } Y_0 = y_0,$$

where B is a standard Brownian motion. As stated in Remark 6.1 (c), the 2nd change-of-variable formula (6.4) and the time-changed Itô formula (6.8) are useful tools in handling a larger class of SDEs of the form

$$dX_t = \rho(t, E_t, X_t)dt + \mu(t, E_t, X_t)dE_t + \sigma(t, E_t, X_t)dB_{E_t} \quad \text{with } X_0 = x_0,$$

where E is a continuous time-change. Note that the measure induced by the sample path $t \mapsto E_t$ is not necessarily absolutely continuous with respect to Lebesgue measure. Hence, the dE_t term appearing above in general cannot be rewritten in terms of dt. For example, if $E = E^\beta$ is an inverse stable subordinator introduced in Section 5.4, then the measure induced by E is not absolutely continuous with respect to Lebesgue measure.

The aim of this section is to make comparisons between the class of classical SDEs and the larger class of SDEs involving time-changed processes. For a general treatment of classical Itô SDEs, see [Ikeda and Watanabe (1989), Karatzas and Shreve (1991)]. Regarding methods for obtaining explicit forms of solutions to classical linear SDEs, consult [Gard (1988), Chap. 4]. Many basic models using classical SDEs are introduced in [Steele (2001)] with an abundance of interpretations and insights. On the other hand, an extensive treatment of the larger class of SDEs involving a time-change is provided in [Kobayashi (2011)].

Let Z be an $(\mathcal{F}_t)$-semimartingale and let E be a continuous $(\mathcal{F}_t)$-time-change. For the purpose of our discussion in this book, we confine our attention to SDEs of the form

$$dX_t = \mu(t, E_t, X_{t-})dE_t + \sigma(t, E_t, X_{t-})dZ_{E_t} \quad \text{with } X_0 = x_0, \qquad (6.17)$$

which is understood in the following integral form:

$$X_t = x_0 + \int_0^t \mu(s, E_s, X_{s-})dE_s + \int_0^t \sigma(s, E_s, X_{s-})dZ_{E_s}, \qquad (6.18)$$

where x_0 is a constant, and μ and σ are defined on $\mathbb{R}_+ \times \mathbb{R}_+ \times \mathbb{R}$ which satisfy the following *Lipschitz condition*: there exists a positive constant C such that

$$|\mu(t, u, x) - \mu(t, u, y)| + \|\sigma(t, u, x) - \sigma(t, u, y)\| \leqslant C|x - y| \qquad (6.19)$$

for all $t, u \in \mathbb{R}_+$ and $x, y \in \mathbb{R}$. For technical reasons, we require the assumption that $(\mu(t, E_t, X_{t-}))$ and $(\sigma(t, E_t, X_{t-}))$ are càglàd for any càdlàg process X. One example of such functions is a 'linear' map $\mu(t, u, x) = \mu_1(t, u) + \mu_2(t, u) \cdot x$, where μ_1, μ_2 are bounded, continuous functions on $\mathbb{R}_+ \times \mathbb{R}_+$. Due to Theorem 7 of [Protter (2004), Chap. V], there exists a unique $(\mathcal{F}_{E_t})$-semimartingale X for which (6.17) holds.

Now that the existence and uniqueness of a solution to an SDE of the form (6.17) is guaranteed, the following two SDEs both make sense:

$$dX_t = \mu(E_t, X_{t-})dE_t + \sigma(E_t, X_{t-})dZ_{E_t} \quad \text{with } X_0 = x_0; \qquad (6.20)$$

$$dY_t = \mu(t, Y_{t-})dt + \sigma(t, Y_{t-})dZ_t \quad \text{with } Y_0 = x_0. \qquad (6.21)$$

Together the change-of-variable formulas (6.1) and (6.4) yield Theorem 6.2, which in turn reveals a close connection between the classical Itô-type SDE (6.21) and our new SDE (6.20).

Theorem 6.2. *(Duality Theorem) Let Z be an $(\mathcal{F}_t)$-semimartingale. Let U and E be a pair satisfying $[[U \longmapsto E]]$ or $[[U \longleftarrow E]]$.*

(1) If a process Y satisfies SDE (6.21), then $X := Y \circ E$ satisfies SDE (6.20).

(2) If a process X satisfies SDE (6.20), then $Y := X \circ U$ satisfies SDE (6.21).

Proof. (1) Suppose Y satisfies SDE (6.21), and let $X := Y \circ E$. Since any process is in synchronization with the continuous $(\mathcal{F}_t)$-time-change E, the 1st change-of-variable formula (6.1) yields

$$X_t = x_0 + \int_0^{E_t} \mu(s, Y_{s-})ds + \int_0^{E_t} \sigma(s, Y_{s-})dZ_s \qquad (6.22)$$

$$= x_0 + \int_0^t \mu(E_s, Y_{E(s)-})dE_s + \int_0^t \sigma(E_s, Y_{E(s)-})dZ_{E_s}.$$

In general, the equality $Y_{E(s)-} = (Y \circ E)_{s-}$ may fail. The failure can occur only when E is constant on some interval $[s - \varepsilon, s]$ with $\varepsilon > 0$. However, the integrators E and $Z \circ E$ on the right hand side of (6.22) are constant on this interval; hence, the difference between the two values $Y_{E(s)-}$ and $(Y \circ E)_{s-}$ does not affect the value of the integrals. Thus, (6.22) can be reexpressed as

$$X_t = x_0 + \int_0^t \mu(E_s, (Y \circ E)_{s-})dE_s + \int_0^t \sigma(E_s, (Y \circ E)_{s-})dZ_{E_s}, \qquad (6.23)$$

thereby yielding SDE (6.20).

(2) Next, suppose X satisfying SDE (6.20) is given. Since U is strictly increasing, $X_{U(s-)-} = (X \circ U)_{s-}$ for any $s > 0$. Again, since any process is in synchronization with the continuous $(\mathcal{F}_t)$-time-change E, the 2nd change-of-variable formula (6.4) applied to the integral form of SDE (6.20) yields

$$X_t = x_0 + \int_0^{E_t} \mu(E_{U(s-)}, X_{U(s-)-})ds + \int_0^{E_t} \sigma(E_{U(s-)}, X_{U(s-)-})dZ_s \qquad (6.24)$$

$$= x_0 + \int_0^{E_t} \mu(s, (X \circ U)_{s-})ds + \int_0^{E_t} \sigma(s, (X \circ U)_{s-})dZ_s.$$

Let $Y := X \circ U$, then (6.24) immediately yields SDE (6.21), which completes the proof. $\square$

Remark 6.2. (a) One may wonder whether the SDE

$$dX_t = \rho(E_t, X_{t-})dt + \mu(E_t, X_{t-})dE_t + \sigma(E_t, X_{t-})dZ_{E_t}$$

can be reduced in the same manner as Theorem 6.2 (2). This is a question of whether the driving process dt can be replaced by dU_{E_t}, which is possible only in very special cases; e.g., if U is continuous or $\rho(E_t, X_{t-})$ vanishes on every nonempty open interval $(U_{\tau-}, U_\tau)$.

(b) Since every Lévy process L is a semimartingale and any inverse stable subordinator E^β of index $\beta \in (0, 1)$ is a continuous time-change, Theorem 6.2 holds for $Z = L$ and $E = E^\beta$. Namely, the duality exists between a classical SDE appearing in (5.17), which is of the form

$$dY_t = b(Y_{t-})dt + \sigma(Y_{t-})dB_t \qquad (6.25)$$

$$+ \int_{|w|<1} H(Y_{t-}, w)\tilde{N}(dt, dw) + \int_{|w|\geqslant 1} K(Y_{t-}, w)N(dt, dw),$$

and a new SDE

$$dX_t = b(X_{t-})dE_t^\beta + \sigma(X_{t-})dB_{E_t^\beta} \tag{6.26}$$
$$+ \int_{|w|<1} H(X_{t-},w)\tilde{N}(dE_t^\beta, dw) + \int_{|w|\geq 1} K(X_{t-},w)N(dE_t^\beta, dw).$$

(c) Detailed analysis of SDEs driven by a time-changed Brownian motion and their concrete examples, including those with linear coefficients, are provided in [Kobayashi (2011)].

6.3 CTRW approximations of time-changed processes in the Skorokhod spaces

By definition, a *continuous time random walk (CTRW)* is a random walk subordinated to a renewal process. Namely, CTRW is defined by a sequence of i.i.d. random variables (or random vectors) called jumps with random arrival times between consecutive jumps. Arrival times are assumed to be positive i.i.d. random variables. Papers [Gorenflo and Mainardi (1998),Meerschaert and Scheffler (2004),Meerschaert and Scheffler (2001),Umarov and Gorenflo (2005b),Umarov and Steinberg (2006)] establish that time-fractional versions of Fokker-Planck-Kolmogorov equations with certain pseudo-differential operators on the right are connected with scaling limit processes of weakly convergent sequences of CTRWs, or more generally, scaling limits of triangular arrays of CTRWs. These limit processes are time-changed Lévy processes, where the time-change arises as the first hitting time of level t (equivalently, the inverse subordinator) for a single stable subordinator. More precisely, the resulting limit stochastic processes have transition probabilities which satisfy associated time-fractional order pseudo-differential equations.

As is noted above, CTRWs are described by two sequences of random variables: one representing the height of the jumps, the other representing the time spent waiting between successive jumps. Consider an example. Let $\{Y_i; i \geq 1\}$ be a sequence of mean zero independent and identically distributed (i.i.d.) random variables in the strict domain of normal attraction of a stable law of index $\alpha \in (1,2]$, which is independent of a sequence of i.i.d. nonnegative random variables $\{J_i; i \geq 1\}$ in the strict domain of normal attraction of a stable law with index $\beta \in (0,1)$; see Section 5.3 for the concept of the domain of attraction of a stable law. One can consider a special case when J_1 has density $f_{J_1}(t) \sim ct^{-1-\beta}$ as $t \to \infty$. Let

$$Y_i^{(c)} = c^{-1/\alpha}Y_i, \quad J_i^{(c)} = c^{-1/\beta}J_i,$$
$$S_n^{(c)} = \sum_{i=1}^n Y_i^{(c)}, \quad \text{and}$$
$$N_t^{(c)} = \max\{n \geq 0 : J_1^{(c)} + \cdots + J_n^{(c)} \leq t\}.$$

It turns out (see below) that the convergence

$$W_t^{(c)} = S_{N_t^{(c)}}^{(c)} = \sum_{i=1}^{N_t^{(c)}} Y_i^{(c)} \to L_{E_t^\beta} \quad \text{as} \quad c \to \infty \tag{6.27}$$

holds in various topologies, where L_t is an α-stable process if $\alpha \in (1,2)$ and a Brownian motion if $\alpha = 2$, and E_t^β is an inverse β-stable subordinator independent of L_t. Thus, in this particular example, the scaling limit (CTRW limit) is a *time-changed Lévy process* $L_{E_t^\beta}$. In this section we discuss convergence in the J_1 and M_1 topologies in the Skorokhod space introduced in Section 5.1 and in the next section in the sense of finite dimensional distributions. As before, the shorthand notations $\xrightarrow{J_1}$ and $\xrightarrow{M_1}$ are used to denote weak convergence on the appropriate Skorokhod spaces in the J_1 and M_1 topologies, respectively.

In the multivariate case, let a sequence of i.i.d. random vectors $\{Y_i \in \mathbb{R}^d : i \geq 1\}$ and a sequence of positive i.i.d. random variables $\{J_i : i \geq 1\}$ be given. In the random walk interpretation the random variable J_i represents the waiting time between the $(i-1)$st step (jump) and the ith step (jump). For each integer $n \geq 1$, the sum

$$S_n = Y_1 + \cdots + Y_n \tag{6.28}$$

is the position after n steps (jumps), while the sum

$$T_n = J_1 + \cdots + J_n \tag{6.29}$$

is the time at which the nth step (jump) occurs. For convenience, we set $S_0 = 0$ and $T_0 = 0$ (though this is not necessary). Then the CTRW process is the stochastic process $W = (W_t)_{t \geq 0}$ defined by

$$W_t = (S \circ N)_t = S_{N_t} = \sum_{i=1}^{N_t} Y_i, \tag{6.30}$$

where

$$N_t = \max\{n \geq 0 : T_n \leq t\} \tag{6.31}$$

represents the number of jumps up to time t. The process N_t is regarded as an inverse to T_n due to the inverse relationship $\{T_n \leq t\} = \{N_t \geq n\}$. If the step lengths (jump heights) and waiting times between consecutive steps (jumps) are independent, then CTRW is called *uncoupled*, otherwise it is called *coupled*.

We are interested in the scaling limit of CTRW processes. Following the discussion given in [Meerschaert and Scheffler (2004)], assume that the random vector Y_i belongs to the strict generalized domain of attraction of a full operator stable law. "Full" means that the measure ν defining the law is not supported on any proper hyperplane of $\mathbb{R}^d$. The law ν being operator stable means that there exist a function $B : \mathbb{R} \to \mathbb{R}^d \times \mathbb{R}^d$ and a $d \times d$ real matrix E such that $B(\kappa)$ is invertible for all $\kappa > 0$, $B(\lambda \kappa)B(\kappa)^{-1} \to \lambda^{-E}$ as $\kappa \to \infty$ for all $\lambda > 0$, and $B(n)S_n \Longrightarrow A$ as $n \to \infty$, where A is a random vector with law ν. Here, $\Longrightarrow$ denotes weak convergence of a sequence of random vectors, λ^{-E} is defined by a power series as

$$\lambda^{-E} = \exp(-E\log\lambda) = \sum_{n=0}^{\infty} \frac{(-\log\lambda)^n}{n!} E^n,$$

and the law ν satisfies $\nu^t(G) = t^E\nu(G) \equiv \nu(t^{-E}G)$ for all Borel sets G in $\mathbb{R}^d$. The matrix E is called the *exponent* of ν and the real parts of eigenvalues of E are in $[1/2, \infty)$. In particular, if $E = (1/\alpha)I$, where $\alpha \in (0, 2]$ and I is the identity matrix, then ν is a stable law with index α. Let

$$S_t = \sum_{i-1}^{\lfloor t \rfloor} Y_i. \tag{6.32}$$

Then by [Meerschaert and Scheffler (2004), Thm. 4.1],

$$B(\kappa)S_{\kappa t} \xrightarrow{J_1} L_t \quad \text{in } \mathbb{D}([0, \infty), \mathbb{R}^d) \text{ as } \kappa \to \infty, \tag{6.33}$$

where $(L_t)_{t \geqslant 0}$ is an operator stable Lévy process with the same exponent E.

Further, assume that the random variable J_i is positive and belongs to the strict domain of attraction of a stable law with index $\beta \in (0, 1)$. Let

$$T_t = \sum_{i=1}^{\lfloor t \rfloor} J_i. \tag{6.34}$$

Then there exists a regularly varying function $b(\kappa)$ with index $-1/\beta$ (i.e. $b(\kappa) = \kappa^{-1/\beta}\ell(\kappa)$ for some slowly varying function $\ell(\kappa)$; see [Bingham et al. (1987), Seneta (1976)]) such that, due to [Meerschaert and Scheffler (2004), Thm. 4.1] (also see [Meerschaert and Scheffler (2008), Example 2.5]),

$$b(\kappa)T_{\kappa t} \xrightarrow{J_1} U_t^{\beta} \quad \text{in } \mathbb{D}([0, \infty), [0, \infty)) \text{ as } \kappa \to \infty,$$

where $(U_t^{\beta})_{t \geqslant 0}$ is a β-stable subordinator. By [Seneta (1976), Property 1.5.5], there exists a regularly varying function $\tilde{b}$ with index β such that

$$b(\tilde{b}(\kappa)) \sim \frac{1}{\kappa} \quad \text{as } \kappa \to \infty.$$

It follows from [Meerschaert and Scheffler (2004), Cor. 3.4] that

$$\tilde{b}(\kappa)^{-1}N_{\kappa t}, \xrightarrow{J_1} E_t^{\beta} \quad \text{in } \mathbb{D}([0, \infty), [0, \infty)) \text{ as } \kappa \to \infty, \tag{6.35}$$

where N_t is defined as in (6.31) and $(E_t^{\beta})_{t \geqslant 0}$ is the inverse of (U_t^{β}).

For this particular CTRW $W_t = S_{N_t}$, the following limit theorem holds.

Theorem 6.3. [Meerschaert and Scheffler (2004), Thm. 4.2] *Let $\{Y_i; i \geqslant 1\}$ be a sequence of random vectors in the strict generalized domain of attraction of a full operator stable law ν with exponent E. Let $\{J_i; i \geqslant 1\}$ be a sequence of positive random variables in the strict domain of attraction of a stable law with index $\beta \in (0, 1)$.*

Suppose that the two sequences are independent. Let B, b and $\tilde{b}$ be the functions given above. For each $\kappa > 0$, define a stochastic process $(W_t^{(\kappa)})_{t \geq 0}$ by

$$W_t^{(\kappa)} = B(\tilde{b}(\kappa))W_{\kappa t} = B(\tilde{b}(\kappa))S_{N_{\kappa t}}, \tag{6.36}$$

where W_t is defined as in (6.30). Then

$$W_t^{(\kappa)} \xrightarrow{M_1} L_{E_t^{\beta}} \quad \text{in } \mathbb{D}([0,\infty),\mathbb{R}^d) \text{ as } \kappa \to \infty, \tag{6.37}$$

where $(L_t)_{t \geq 0}$ is an operator stable Lévy process with exponent E and $(E_t^{\beta})_{t \geq 0}$ is an inverse β-stable subordinator independent of (L_t).

The proof of Theorem 6.3 is delayed until after Theorem 6.5. The proof provided in [Meerschaert and Scheffler (2004)] uses a generic theorem in [Whitt (2002)] (Theorem 6.4 below) on convergence of sequences of composite stochastic processes in the M_1 topology. Applying a generic theorem on convergence of sequences of composite stochastic processes in the J_1 topology, formulated in [Silvestrov (2004)] (Theorem 6.5 below), we prove that the convergence of the CTRW processes given in (6.37) holds in the stronger J_1 topology.

The stochastic process $W_t^{(\kappa)}$ defined in (6.36) can be written as the composition of the two càdlàg stochastic processes $S_t^{(\kappa)} \equiv B(\tilde{b}(\kappa))S_{\tilde{b}(\kappa)t}$ and $N_t^{(\kappa)} \equiv \tilde{b}(\kappa)^{-1}N_{\kappa t}$. More generally, scaled CTRW processes are given in the form

$$W_t^{(\kappa)} = (S^{(\kappa)} \circ N^{(\kappa)})_t \tag{6.38}$$

for some scaled processes $S_t^{(\kappa)}$ and $N_t^{(\kappa)}$. The following two general theorems on convergence of sequences of composite stochastic processes in the Skorokhod space supply sufficient conditions for convergence of CTRWs in the J_1 and M_1 topologies. We recall that the set $Disc(x)$ for a càdlàg function $x(t)$ means the set of its discontinuity points $\{t : x(t-) - x(t) \neq 0\}$. The first theorem is a stochastic process version of [Whitt (2002), Thm. 13.2.4].

Theorem 6.4. [Whitt (2002), Thm. 13.2.4.] *Let X_n, $n \geq 1$, and X_0 be càdlàg d-dimensional stochastic processes and V_n, $n \geq 1$, and V_0 be non-negative and non-decreasing one-dimensional càdlàg stochastic processes such that $(X_n, V_n) \xrightarrow{M_1} (X_0, V_0)$ in $\mathbb{D}([0,\infty),\mathbb{R}^d) \times \mathbb{D}([0,\infty),\mathbb{R})$. If*

(1) *a.s., $V_0(t)$ is continuous and strictly increasing at t whenever $V_0(t) \in Disc(X_0)$;*

(2) *a.s., X_0 is monotone on $[V_0(t-), V_0(t)]$ and $V_0(t-)$, $V_0(t) \notin Disc(X_0)$ whenever $t \in Disc(V_0)$,*

then $X_n \circ V_n \xrightarrow{M_1} X_0 \circ V_0$ in $\mathbb{D}([0,\infty),\mathbb{R}^d)$ as $n \to \infty$.

The second theorem concerns convergence of sequences of composite stochastic processes in the J_1 topology, which is stronger than the M_1 topology. The theorem appears in [Silvestrov (2004)] and requires us to recall the J_1 compactness condition appearing in Theorem 5.1.

Theorem 6.5. *[Silvestrov (2004), Thm. 3.4.2] Let X_ε, $\varepsilon > 0$, and X_0 be càdlàg d-dimensional stochastic processes, and V_ε, $\varepsilon > 0$, and V_0 be non-negative and non-decreasing one-dimensional càdlàg stochastic processes. Assume that the following conditions hold:*

(1) $(X_\varepsilon(t), V_\varepsilon(s))_{(t,s)\in T\times S}$ *converges weakly to* $(X_0(t), V_0(s))_{(t,s)\in T\times S}$ *as $\varepsilon \to 0$, where T and S are dense subsets of $[0,\infty)$ containing the point 0;*

(2) $\lim\limits_{c\to 0} \limsup\limits_{\varepsilon\to 0} \mathbb{P}\left[\omega_J(X_\varepsilon(\cdot), c, T) > \delta\right] = 0$ *for all $\delta > 0$ and $T > 0$;*

(3) V_0 is an a.s. continuous process;

(4) $\mathbb{P}\left[V_0(t') = V_0(t'') \in Disc(X_0)\right] = 0$ *for $0 \leqslant t' < t'' < \infty$;*

(5) $\mathbb{P}\left[V_0(0) \in Disc(X_0)\right] = 0$.

Then $X_\varepsilon \circ V_\varepsilon \xrightarrow{J_1} X_0 \circ V_0$ *in $\mathbb{D}([0,\infty), \mathbb{R}^d)$ as $\varepsilon \to 0$.*

Theorem 6.4 applied to $X_n(t) = B(\tilde{b}(n))S_{\tilde{b}(n)t}$ and $V_n(t) = \tilde{b}(n)^{-1}N_{nt}$ enables us to prove Theorem 6.3 in a simple manner as follows.

Proof of Theorem 6.3. Since $\{J_i\}$ and $\{Y_i\}$ are independent, so are $(S_t)_{t\geqslant 0}$ and $(N_t)_{t\geqslant 0}$. Hence, by (6.33) and (6.35),

$$\left(S^{(\tilde{b}(\kappa))}_{\tilde{b}(\kappa)t}, \tilde{b}(\kappa)^{-1}N_{\kappa t}\right) \xrightarrow{J_1} (L_t, E_t^\beta)$$

in $\mathbb{D}([0,\infty), \mathbb{R}^d) \times \mathbb{D}([0,\infty), [0,\infty))$ as $\kappa \to \infty$. Since the M_1 topology is weaker than the J_1 topology, this convergence also holds in the M_1 topology. Therefore, the desired conclusion follows once we verify the conditions of Theorem 6.4 with $X_0(t) = L_t$ and $V_0(t) = E_t^\beta$. Note that E_t^β is continuous a.s. since U_t^β is strictly increasing a.s., and hence, condition (2) is vacuously satisfied. Also, since the independent Lévy processes L_t and U_t^β have no simultaneous jumps a.s., condition (1) is also satisfied. This completes the proof. □

Theorem 6.5 applied to $X_\varepsilon(t) = B(\tilde{b}(1/\varepsilon))S_{\tilde{b}(1/\varepsilon)t}$ and $V_\varepsilon(t) = \tilde{b}(1/\varepsilon)^{-1}N_{t/\varepsilon}$ with $\varepsilon = 1/\kappa$ enables us to prove the J_1 convergence of the CTRW $W_t^{(\kappa)}$ defined in (6.36).

Theorem 6.6. *Under the conditions of Theorem 6.3, $W_t^{(\kappa)} \xrightarrow{J_1} L_{E_t}$ in $\mathbb{D}([0,\infty), \mathbb{R}^d)$ as $\kappa \to \infty$.*

Proof. Let $X_\varepsilon(t) = B(\tilde{b}(1/\epsilon))S_{\tilde{b}(1/\epsilon)t}$ and $V_\varepsilon(t) = \tilde{b}(1/\epsilon)^{-1}N_{t/\varepsilon}$ in which $\varepsilon = 1/\kappa$. Then the composition process $(X_\varepsilon \circ V_\varepsilon)(t)$ represents the CTRW process $W_t^{(\kappa)}$. Therefore, it suffices to check each condition of Silvestrov's Theorem 6.5 with $X_0(t) = L_t$ and $V_0(t) = E_t^\beta$.

Condition (1) of this theorem follows from (6.33) and (6.35) together with the independence between X_ε and V_ε. The fact that $X_\varepsilon(t)$ satisfies condition (2) is verified in the proof of Theorem 4.1 of [Meerschaert and Scheffler (2004)]. Since the stable subordinator U_t^β is strictly increasing a.s., its inverse E_t^β is continuous a.s., and hence, condition (3) holds. Note that the independent Lévy processes L_t and U_t^β have no simultaneous jumps a.s. This implies that a.s., $V_0(t)$ is strictly increasing at t whenever $V_0(t) \in Disc(X_0)$, which is equivalent to condition (4). Moreover, since $E_0^\beta = 0$, condition (5) is reduced to $\mathbb{P}[0 \in Disc(X_0)] = 0$, or equivalently, a.s. L_t has no discontinuity at $t = 0$, which follows from the assumption that L_t has right-continuous paths.

Hence, in accordance with Theorem 6.5, the composite process $X_\varepsilon \circ V_\varepsilon$ converges to the composite process $X_0 \circ V_0$ as $\varepsilon \to 0$ in the J_1 topology in $\mathbb{D}([0,\infty), \mathbb{R}^d)$, or equivalently, the CTRW process $W^{(\kappa)}$ converges to the process $L \circ E^\beta$ as $\kappa \to \infty$ in the J_1 topology in $\mathbb{D}([0,\infty), \mathbb{R}^d)$. □

The properties of the limiting stochastic process in the above theorems are collected in the following statement.

Theorem 6.7. [Meerschaert and Scheffler (2004)] *The limiting stochastic process $M_t = L_{E_t^\beta}$ obtained in Theorems 6.3 and 6.6 possesses the following properties:*

(1) *M_t is operator self-similar with exponent βE with $\beta \in (0,1)$ and matrix E, that is, for all $c > 0$,*

$$\{M_{ct} : t \geqslant 0\} = \{c^{\beta E} M_t : t \geqslant 0\}$$

holds with equality in the sense of finite-dimensional distributions;
(2) *M_t has no stationary increments;*
(3) *M_t is not operator stable for any $t > 0$;*
(4) *M_t is non-Markovian.*

An important question related to fractional FPK equations is: What connection is there between the CTRW limiting processes and the fractional order differential equations? When the scaling limit of CTRWs is $X = L \circ E^\beta$, where L is an α-stable Lévy process and E^β is an independent time-change given by an inverse β-stable subordinator as in (6.27), the transition probability $p^X(t, x)$ of X satisfies the Cauchy problem

$$D_*^\beta p^X(t, x) = \mathcal{A}(D_x) p^X(t, x), \quad p^X(0, x) = \delta_0(x), \tag{6.39}$$

where D_*^β is the fractional derivative of order β in the sense of Caputo–Djrbashian, $\mathcal{A}(D_x)$ is the pseudo-differential operator with a constant symbol $\Psi(\xi)$, and $\delta_0(x)$ is the Dirac delta function with mass on 0. Alternatively, (6.39) can be represented in the form

$$D_+^\beta p^X(t, x) = \mathcal{A}(D_x) p^X(t, x) + \frac{t^{-\beta}}{\Gamma(1-\beta)} \delta_0(x),$$

with the Riemann–Liouville fractional derivative D_+^β. In general (see papers [Baeumer et al. (2005a), Bazhlekova (2000)]), if $\mathcal{A}$ is the infinitesimal generator of a strongly continuous semigroup in a Banach space and $v(t)$ is a vector-function whose Laplace transform is given by

$$\tilde{v}(s) = s^{\beta-1} R(\mathcal{A}, s^\beta)\varphi,$$

where $R(\mathcal{A}, s)$ is the resolvent operator of $\mathcal{A}$ and φ is in the domain of $\mathcal{A}$, then $v(t)$ solves the fractional order abstract Cauchy problem $D_*^\beta v(t) = \mathcal{A}v(t)$, $v(0) = \varphi$. As we will see in the next chapter, if such a semigroup corresponds to a stochastic process which solves an SDE driven by a Lévy process, then the operator $\mathcal{A}$ on the right hand side of this equation is given by a pseudo-differential operator of the form $\mathcal{A}(x, \mathbf{D}_x)$ given in (5.22) with the symbol $\Psi(x, \xi)$ defined in (5.21).

6.4   CTRW approximations of time-changed processes in the sense of finite-dimensional distributions

Below we examine the connection of CTRW limits with fractional order differential equations. As is suggested by condition (C) in Theorems 5.1 and 5.2, weak convergence of CTRWs in the sense of finite dimensional distributions is an important step in the study of convergence of the CTRWs in the J_1 and M_1 topologies of the Skorokhod space. Weak convergence of CTRWs in the sense of finite dimensional distributions is studied in the works [Gorenflo and Mainardi (1998), Gorenflo and Mainardi (1999), Gorenflo and Mainardi (1999), Umarov and Gorenflo (2005b), Liu et al. (2005), Umarov and Steinberg (2006), Umarov (2015a)]. Below we briefly discuss some of these results.

The paper [Gorenflo and Mainardi (1998)] contains a construction of several versions of CTRWs all converging weakly to stochastic processes whose density functions satisfy the associated fractional order pseudo-differential equations. One example is the Gillis–Weiss random walk, which is constructed as follows. Let $0 < \alpha \leqslant 2$, and $\{X_i; i \geqslant 1\}$ be a sequence of i.i.d. discrete random variables with values in the set of integers $\mathbb{Z}$ with the probability mass function

$$\mathbb{P}(X_1 = k) = p_k = \begin{cases} \lambda |k|^{\alpha+1}, & \text{if } k \neq 0, \\ 1 - 2\lambda\zeta(\alpha+1), & \text{if } k = 0, \end{cases} \tag{6.40}$$

where $\zeta(z)$ is the Riemann zeta function defined as $\zeta(z) = \sum_{k=1}^\infty k^{-z}$, $Re(z) > 1$, and $\lambda > 0$ satisfies the condition $\lambda \leqslant (2\zeta(\alpha+1))^{-1}$.

Theorem 6.8. [Gorenflo and Mainardi (1998)] *Let $0 < \alpha \leqslant 2$ and $\{X_i; i \geqslant 1\}$ be a sequence of i.i.d. random variables with the probability mass function defined in (6.40). For $h > 0$, set the scaling relation*

$$\tau = \begin{cases} \mu h^\alpha, & \text{if } 0 < \alpha < 2, \\ \lambda h^2 \ln \frac{1}{h}, & \text{if } \alpha = 2, \end{cases}$$

where $0 < \mu < \pi \left(2\Gamma(\alpha+1)\zeta(\alpha+1)\sin(\alpha\pi/2)\right)^{-1}$. Then as $N = \lfloor \frac{t}{\tau} \rfloor + 1 \to \infty$ with $t \geqslant 0$, the CTRWs $S_N = hX_1 + ... + hX_N$ converge weakly in the sense of finite-dimensional distributions to a stochastic process whose probability density function *is the solution to equation*

$$\frac{\partial u(t,x)}{\partial t} = \mathbb{D}_0^\alpha u(t,x), \quad t > 0, \, x \in \mathbb{R}, \tag{6.41}$$

with the initial condition $u(0,x) = \delta_0(x)$. *Here* $\mathbb{D}_0^\alpha$ *is the pseudo-differential operator defined in* (3.44) *in Section 3.7 if* $0 < \alpha < 2$; *it is the Laplace operator* Δ *if* $\alpha = 2$.

Note that the pseudo-differential operator $\mathbb{D}_0^\alpha$ has symbol $\sigma_{\mathbb{D}_0^\alpha}(\xi) = -|\xi|^\alpha$. Theorem 6.8, as well as other theorems related to CTRW limits such as the Grünwald–Letnikov, globally binomial, and Chechkin–Gonchar random walks, studied in [Gorenflo and Mainardi (1998)], are proved in the one-dimensional case. A multidimensional version of the Gillis–Weiss random walk and its generalizations to a sequence of i.i.d. random vectors $\{X_i; i \geqslant 1\}$ which are mixed stables with a mixing measure ρ are studied in the works [Umarov and Gorenflo (2005b), Umarov and Steinberg (2006)].

Theorem 6.9. [Umarov and Steinberg (2006)] *For* $h > 0$ *and* $\tau > 0$, *let* X *be a random vector with the probability mass function* $p_k = \mathbb{P}(X = k)$, $k \in \mathbb{Z}^d$, *defined by*

$$p_k = \begin{cases} 1 - \tau \sum_{m \neq 0} \frac{Q_m(h)}{|m|^d} & \text{if } k = 0 \\ \tau \frac{Q_k(h)}{|k|^d} & \text{if } k \neq 0, \end{cases}$$

where

$$Q_m(h) = \int_0^2 \frac{B_{d,\alpha} d\rho(\alpha)}{h^\alpha |m|^\alpha}, \tag{6.42}$$

with $B_{d,\alpha}$ *defined as in equation* (3.45) *and* ρ *is a positive Borel measure defined on* $(0,2)$. *Assume the condition*

$$\tau \leqslant \frac{1}{2} \left(\sum_{m \neq 0} \frac{Q_m(h)}{|m|^d} \right)^{-1}.$$

Let X_j, $j = 1, 2, \ldots$, *be i.i.d. random vectors distributed as* X. *Then as* $N = \lfloor \frac{t}{\tau} \rfloor + 1 \to \infty$ *with* $t \geqslant 0$, *the CTRWs* $S_N = hX_1 + \ldots + hX_N$ *converge weakly in the sense of finite-dimensional distributions to a stochastic process whose probability density function is the fundamental solution of the distributed space fractional order differential equation*

$$\frac{\partial u(t,x)}{\partial t} = \Psi(D)u(t,x), \quad t > 0, \, x \in \mathbb{R}^d,$$

with the initial condition $u(0,x) = \delta_0(x)$. *Here* $\Psi(D)$ *is the pseudo-differential operator with the symbol*

$$\psi(\xi) = -\int_0^2 |\xi|^\alpha d\rho(\alpha). \tag{6.43}$$

A construction of CTRWs associated with time-fractional pseudo-differential equations requires the so-called non-Markovian transition probabilities in addition to the Markovian transition probabilities p_k. Two different versions of such CTRWs were suggested in the works [Gorenflo and Mainardi (2001), Anh and McVinish (2003)]. In these papers convergence of the CTRW to the limiting process is established numerically. An analytic proof of convergence in the one-dimensional case is established in [Abdel-Rehim (2013)] and in the multidimensional case in [Umarov (2015a)] using a different method.

Theorem 6.10. [Umarov (2015a)] *Let $0 < \beta \leq 1$. For $h > 0$ and $\tau > 0$, let $Y_j \in \mathbb{Z}^d$, $j \geq 1$, be identically distributed random vectors with the non-Markovian probabilities*

$$c_\ell = (-1)^{\ell+1} \binom{\beta}{\ell} = \left| \binom{\beta}{\ell} \right|, \quad \ell = 1, \dots, n,$$

$$\gamma_n = \sum_{\ell=0}^{n} (-1)^\ell \binom{\beta}{\ell},$$

and Markovian transition probabilities

$$p_k = \begin{cases} c_1 - \tau^\beta \sum_{m \neq 0} \frac{Q_m(h)}{|m|^d} & \text{if } k = 0 \\ \tau^\beta \frac{Q_k(h)}{|k|^d} & \text{if } k \neq 0, \end{cases}$$

where $Q_m(h)$ is defined in (6.42). Assume that

$$\tau \leq \left(\frac{\beta}{Q(h)} \right)^{\frac{1}{\beta}}. \tag{6.44}$$

Then as $N = \lfloor \frac{t}{\tau} \rfloor + 1 \to \infty$ with $t \geq 0$, the CTRWs $S_N = hX_1 + \dots + hX_N$ converge weakly in the sense of finite-dimensional distributions to a time-changed stochastic process whose probability density function is the solution to the time-fractional pseudo-differential equation

$$D_*^\beta u(t, x) = \Psi(D)u(t, x), \quad t > 0, \ x \in \mathbb{R}^d,$$

with the initial condition $u(0, x) = \delta_0(x)$. Here D_*^β is the fractional derivative of order β in the sense of Caputo–Djrbashian (see Section 3.2) and $\Psi(D)$ is the pseudo-differential operator with the symbol in equation (6.43).

The role of the non-Markovian probabilities in the above theorem is that the probability q_j^{n+1} of the constructed random walk taking on the value $hj = (hj_1, \dots, hj_d)$ at time instant t_{n+1} is recursively calculated by

$$q_j^{n+1} = \gamma_n q_j^0 + \sum_{\ell=1}^{n-1} c_{n-\ell+1} q_j^\ell + \left(c_1 - \tau^\beta Q_0(h) \right) q_j^n + \sum_{m \neq 0} p_m q_{j-m}^n.$$

See [Umarov (2015b)] for a CTRW construction associated with more general time-fractional distributed pseudo-differential equations.

6.5   Approximations of stochastic integrals driven by time-changed processes

Another important but less studied case of CTRW approximation is the approximation of stochastic integrals driven by scaling limits of CTRWs. We can only cite a few papers on this topic. Weak convergence of stochastic integrals driven by scaled CTRWs to a stochastic integral driven by the limit of the scaled CTRWs is first studied in [Burr (2011)]. The results in [Burr (2011)] are derived from some more general results on weak convergence of stochastic integrals in the J_1 topology in $\mathbb{D}([0,\infty), \mathbb{R}^d)$ from [Kurtz and Protter (1996),Kurtz and Protter (1991a)]. Thus, we begin by reviewing the general set-up and accompanying general results. For simplicity, we deal with the one-dimensional case.

Let $\Theta^n = (\Omega^n, \mathcal{F}^n, (\mathcal{F}_t^n)_{t \geqslant 0}, \mathbb{P}^n)$ be a sequence of filtered probability spaces satisfying the usual conditions (i.e. $\mathcal{F}_0^n$ contains all null sets and and the filtration $(\mathcal{F}_t^n)$ is right-continuous). Assume

(1) $\{Z_t^{(n)}, n \geqslant 1\}$ is a sequence of $\mathcal{F}_t^n$-semimartingales in $\mathbb{D}([0,\infty), \mathbb{R})$;
(2) $\{H_t^{(n)}, n \geqslant 1\}$ is a sequence of $\mathcal{F}_t^n$-adapted processes in $\mathbb{D}([0,\infty), \mathbb{R})$;
(3) $Z_t^{(n)} \xrightarrow{J_1} Z_t$ in $\mathbb{D}([0,\infty), \mathbb{R})$.

The paper [Kurtz and Protter (1996)] defines a sequence of integrators in the approximating integrals as *good* integrators as follows. A sequence of integrators $\{Z_t^{(n)}, n \geqslant 1\}$ is called *good* if for any sequence of integrands $\{H_t^{(n)}, n \geqslant 1\}$ in $\mathbb{D}([0,\infty), \mathbb{R})$ defined on $\{\Theta^n, n \geqslant 1\}$ such that $(H^{(n)}, Z^{(n)}) \xrightarrow{J_1} (H, Z)$ in $\mathbb{D}([0,\infty), \mathbb{R}^2)$ as $n \to \infty$, there exists a filtration $(\mathcal{F}_t)$ such that Z is an $\mathcal{F}_t$-semimartingale and $\int H_{s-}^{(n)} dZ_s^{(n)} \xrightarrow{J_1} \int H_{s-} dZ_s$ in $\mathbb{D}([0,\infty), \mathbb{R})$.

Remark 6.3. Convergence of the joint distributions above occurs in $\mathbb{D}([0,\infty), \mathbb{R}^2)$ rather than $\mathbb{D}([0,\infty), \mathbb{R}) \times \mathbb{D}([0,\infty), \mathbb{R})$. Basically, this requires the change in time in the definition of J_1 convergence to be the same in each component. It is needed to insure the stochastic integrals will be in $\mathbb{D}([0,\infty), \mathbb{R})$.

The objective now is to identify *good* integrators. The above sequence of integrals have values in the metric space, $\mathbb{D}([0,\infty), \mathbb{R})$. Weak convergence in a metric space requires tightness of the laws of the convergent sequences. The appropriate notion of tightness for the laws of the sequence of integrals $\{\int H_{s-}^{(n)} dZ_s^{(n)}, n \geqslant 1\}$ is a condition on the integrators introduced in [Jakubowski et al. (1989)], called the UT condition. A sequence of integrators $\{Z_t^{(n)}, n \geqslant 1\}$ defined on $\{\Theta^n, n \geqslant 1\}$ is called *uniformly tight (UT)* if for $\mathbf{S}^n$, the collection of piecewise constant predictable processes on Θ^n, and each $t > 0$, the set $\{\int_0^t H_{s-}^{(n)} dZ_s^{(n)} : H^{(n)} \in \mathbf{S}^n, |H^{(n)}| \leqslant 1, n \geqslant 1\}$ is stochastically bounded uniformly in n, i.e., for each $t > 0$ and $\varepsilon > 0$, there exists a $C < \infty$ such that

$$\sup \left\{ P^n \left(\left| \int_0^t H_{s-}^{(n)} dZ_s^{(n)} \right| > C \right) : n \geqslant 1, H^{(n)} \in \mathbf{S}^n, \sup_{0 \leqslant s \leqslant t} |H^{(n)}(s)| \leqslant 1 \right\} < \varepsilon.$$

Kurtz and Protter obtained the following two important results:

Theorem 6.11. [Kurtz and Protter (1991a)] *If the joint convergence* $(H^{(n)}, Z^{(n)}) \xrightarrow{J_1} (H, Z)$ *in* $\mathbb{D}([0, \infty), \mathbb{R}^2)$ *holds and if the sequence* $\{Z_t^{(n)}, n \geqslant 1\}$ *satisfies the UT condition, then the sequence* $\{Z_t^{(n)}, n \geqslant 1\}$ *is good, i.e. there exists a filtration* $(\mathcal{F}_t)$ *such that* Z *is an* $\mathcal{F}_t$*-semimartingale and* $\int H_{s-}^{(n)} dZ_s^{(n)} \xrightarrow{J_1} \int H_{s-} dZ_s$ *in* $\mathbb{D}([0, \infty), \mathbb{R})$.

Theorem 6.12. [Kurtz and Protter (1996)] *If both* $Z^{(n)} \xrightarrow{J_1} Z$ *in* $\mathbb{D}([0, \infty), \mathbb{R})$ *and the sequence* $\{Z^{(n)}, n \geqslant 1\}$ *is good, then* $\{Z^{(n)}, n \geqslant 1\}$ *satisfies the UT condition.*

Combining the above two theorems yields the following:

Corollary 6.1. *If* $(H^{(n)}, Z^{(n)}) \xrightarrow{J_1} (H, Z)$ *in* $\mathbb{D}([0, \infty), \mathbb{R}^2)$, *then* $\{Z_t^{(n)}, n \geqslant 1\}$ *is good if and only if* $\{Z_t^{(n)}, n \geqslant 1\}$ *satisfies the UT condition.*

Unfortunately, the definition of UT is difficult to check directly. In cases when $\{Z_t^{(n)}, n \geqslant 1\}$ is a sequence of local martingales, it is usually easier to verify the following condition which implies the UT condition.

Proposition 6.1. [Jakubowski et al. (1989),Kurtz and Protter (1996)] *If* $\{Z_t^{(n)}, n \geqslant 1\}$ *is a sequence of local martingales and if for each* $t < \infty$,

$$\sup_{n \geqslant 1} \mathbb{E}^n \Big[\sup_{0 \leqslant s \leqslant t} |\Delta Z_s^{(n)}| \Big] < \infty,$$

where $\Delta Z_s^{(n)} = X_s - X_{s-}$ *and* $\mathbb{E}^n$ *denotes the expectation under* $\mathbb{P}^n$, *then* $\{Z_t^{(n)}, n \geqslant 1\}$ *satisfies the UT condition.*

Turning now to the CTRW approximation of a stochastic integral driven by a martingale M, let $Z_t^{(n)} = W_t^{(n)}$ be a scaled CTRW and $Z_t = M_t$ be the CTRW limit in the J_1 topology in $\mathbb{D}([0, \infty), \mathbb{R})$. [Burr (2011)] has verified both the conditions of the Theorem 6.11 and, via Proposition 6.1, the UT condition to obtain the following result.

Theorem 6.13. [Burr (2011)] *Let* $\{Y_i; i \geqslant 1\}$ *be a sequence of i.i.d. mean 0 random variables with* $E|Y_i|^2 = c^2 \in (0, \infty)$. *Let* $\{J_i; i \geqslant 1\}$ *be a sequence of i.i.d. strictly* β*-stable random variables,* $\beta \in (0, 1)$. *Assume that the two sequences are independent. Let*

$$W_t^{(n)} = \frac{1}{cn^{\beta/2}} \sum_{i=1}^{N_{nt}} Y_i,$$

where $N_t = \max\{n \geqslant 0 : J_1 + \cdots + J_n \leqslant t\}$. *Then* $W^{(n)} \xrightarrow{J_1} B \circ E$ *in* $\mathbb{D}([0, \infty), \mathbb{R})$ *as* $n \to \infty$, *where* B *is a Brownian motion and* E *is an independent inverse* β*-stable subordinator. Moreover, if* $(H^{(n)}, W^{(n)}) \xrightarrow{J_1} (H, B \circ E)$ *in* $\mathbb{D}([0, \infty), \mathbb{R}^2)$, *then* there exists a filtration $(\mathcal{F}_t)$ such that $B \circ E$ is an $(\mathcal{F}_t)$-semimartingale, H is an $(\mathcal{F}_t)$-adapted càdlàg process, and as $n \to \infty$,

$$\int_0^t H_{s-}^{(n)} dW_s^{(n)} \xrightarrow{J_1} \int_0^t H_{s-} dB_{E_s}, \quad in \; \mathbb{D}([0,\infty), \mathbb{R}). \tag{6.45}$$

Remark 6.4. Theorem 6.13 can be applied for modeling and simulation of stochastic processes serving as solutions to SDEs driven by time-changed Brownian motion. In [Burr (2011)] some of its applications are provided. In order to expand the scope of applications to the case of SDEs driven by time-changed processes to be discussed in Chapter 7, one needs to generalize Theorem 6.13 to the case of a sequence $\{Y_i; i \geqslant 1\}$ of i.i.d. random variables belonging to the strict domain of attraction of operator stable laws. Scalas and Viles [Scalas and Viles (2013)] studied a particular case of approximation of stochastic integrals driven by a time-changed symmetric α-stable Lévy process. In the general case, however, a challenging issue arises connected with verification of the UT condition of the CTRW processes. To discuss this issue, suppose that a sequence of integrands $\{H^{(n)}; n \geqslant 1\}$ and a càdlàg process H satisfy the following conditions:

(1) $(H_t^{(n)})_{t \in S}$ converges weakly to $(H_t)_{t \in S}$ as $n \to \infty$, where S is a dense subset of $[0,\infty)$ containing the point 0;
(2) For $H_t^{(n)}$ the J_1-compactness holds:

$$\lim_{c \to 0} \limsup_{n \to \infty} \; \mathbb{P}\{\omega_J(H^{(n)}, c, T) > \delta\} = 0 \quad \text{for all} \;\; \delta, T > 0.$$

Due to Theorem 5.1, these two conditions imply the convergence $H_t^{(n)} \to H_t$ in the J_1 topology in $\mathbb{D}([0,\infty), \mathbb{R})$. Further, assume that the two sequences of i.i.d. random variables $Y_i^{(n)}$ and $J_i^{(n)}$ satisfy the conditions of Theorem 6.3. Then by Theorem 6.6, the CTRW process $W_t^{(n)}$ converges to the time-changed process $L_{E_t^\beta}$ in the J_1 topology in $\mathbb{D}([0,\infty), \mathbb{R})$. Moreover, due to Proposition 5.3, the sequence of pairs $(H_t^{(n)}, W_t^{(n)})$ converges to the pair $(H_t, L_{E_t^\beta})$ in the J_1 topology in $\mathbb{D}([0,\infty), \mathbb{R}^2)$. Therefore, the first condition in Theorem 6.11 is verified. Now the key issue is to prove the UT condition for the CTRW $W_t^{(n)}$. To the best knowledge of the authors, this fact in the general case has not yet been established.

6.6 Numerical approximations of SDEs driven by a time-changed Brownian motion

This section proposes a discretization scheme for SDEs of the form

$$dX_t = b(E_t, X_t)dE_t + \sigma(E_t, X_t)dB_{E_t} \quad \text{with} \;\; X_0 = x_0, \tag{6.46}$$

where $x_0 \in \mathbb{R}^d$ is a non-random constant, and $b(t,x) : [0,\infty) \times \mathbb{R}^d \to \mathbb{R}^d$ and $\sigma(t,x) : [0,\infty) \times \mathbb{R}^d \to \mathbb{R}^{d \times m}$ are measurable functions for which there is a positive constant K such that

$$|b(t,x) - b(t,y)| + |\sigma(t,x) - \sigma(t,y)| \leqslant K|x - y|, \tag{6.47}$$

$$|b(t,x)| + |\sigma(t,x)| \leqslant K(1 + |x|), \tag{6.48}$$

$$|b(s,x) - b(t,x)| + |\sigma(s,x) - \sigma(t,x)| \leqslant K(1 + |x|)|s - t| \tag{6.49}$$

for all $x, y \in \mathbb{R}^d$ and $s, t \geqslant 0$, where $|\cdot|$ denotes the Euclidean norms of appropriate dimensions. As usual, B is a Brownian motion and E is an independent time-change given by the inverse of a subordinator U with infinite Lévy measure. U has strictly increasing paths with infinitely many jumps (see e.g. [Sato (1999), Thm. 21.3]), which implies that E has continuous, nondecreasing paths. The initial value x_0 is taken to be a non-random constant only for simplicity of discussions. Since (E_t) and (B_{E_t}) are non-Markovian and do not have independent or stationary increments, it is difficult to simulate sample paths of the solution (X_t) to SDE (6.46) via direct applications of well-known approximation schemes such as the Euler–Maruyama scheme (see [Kloeden and Platen (1999)]).

The approximation scheme to be presented in this section appears in [Jum and Kobayashi (2016)] and extends a scheme discussed in Section III of [Gajda and Magdziarz (2010)] to SDEs of the above form with *general* time-dependent coefficients and time-changes. In [Jum and Kobayashi (2016)], the authors further prove that the approximation process converges to the exact solution of the above SDE both strongly and weakly. They also investigated the respective orders of convergence, which is a very important issue in applications.

Since the time-change E is continuous, one can employ the duality theorem (Theorem 6.2) to connect SDE (6.46) with the classical Itô SDE

$$dY_t = b(t, Y_t)dt + \sigma(t, Y_t)dB_t \quad \text{with} \quad Y_0 = x_0. \tag{6.50}$$

The solution Y to this SDE and the solution X to SDE (6.46) are connected by the relationship $X_t = Y_{E_t}$.

The discretization scheme for SDE (6.46) on a fixed interval $[0, T]$ is two-fold; namely, to apply the Euler–Maruyama scheme to SDE (6.50) to construct a process $Y^\delta = (Y_t^\delta)_{t \geqslant 0}$ approximating the solution Y (see (6.51)–(6.52) below), and to approximate the time-change E by a process $E^\delta = (E_t^\delta)_{t \in [0,T]}$ to be defined in (6.54) (which was introduced in [Magdziarz (2009a), Magdziarz (2009b)]). Here, $\delta \in (0, 1)$ denotes the equidistant step size to be taken in the discretization scheme. The duality theorem suggests the use of the process $X^\delta = (X_t^\delta)_{t \in [0,T]}$ defined by

$$X_t^\delta := Y_{E_t^\delta}^\delta$$

as a process approximating the solution X of SDE (6.46). However, to guarantee the reliability of our approximation scheme, we must carefully analyze two different errors—one generated by the Euler–Maruyama scheme and the other due to the approximation of the time-change.

Here, we will present how to construct Y^δ and E^δ and state pertinent convergence results without proofs. For details, the reader is referred to [Jum and Kobayashi (2016)].

Let $\delta \in (0, 1)$ be a fixed number, serving as an equidistant step size of the approximation. We apply the Euler–Maruyama scheme to SDE (6.50) on the positive real line $[0, \infty)$ by choosing discretization times $\tau_n := n\delta$, $n = 0, 1, 2, \ldots$ and then setting

$$Y_0^\delta := x_0, \quad Y_{\tau_{n+1}}^\delta := Y_{\tau_n}^\delta + b(\tau_n, Y_{\tau_n}^\delta)(\tau_{n+1} - \tau_n) \tag{6.51}$$
$$+ \sigma(\tau_n, Y_{\tau_n}^\delta)(B_{\tau_{n+1}} - B_{\tau_n})$$

for $n = 0, 1, 2, \ldots$. A continuous-time process $(Y_t^\delta)_{t \geq 0}$ is defined by continuously interpolating the discrete-time process $(Y_{\tau_n}^\delta)_{n=0,1,2,\ldots}$ by

$$Y_t^\delta := Y_{\tau_n}^\delta + b(\tau_n, Y_{\tau_n}^\delta)(t - \tau_n) \tag{6.52}$$
$$+ \sigma(\tau_n, Y_{\tau_n}^\delta)(B_t - B_{\tau_n}) \text{ whenever } t \in [\tau_n, \tau_{n+1}].$$

The interpolation is for a theoretical purpose only and the information of the interpolated values is not used for simulation of sample paths of the solution X of SDE (6.46).

To approximate the time-change E on a fixed interval $[0, T]$, we follow an idea presented in [Gajda and Magdziarz (2010)]. First, simulate a sample path of the subordinator U, which has independent and stationary increments, by setting $U_0 = 0$ and then following the rule $U_{i\delta} := U_{(i-1)\delta} + Z_i$, $i = 1, 2, 3, \ldots$, where $\{Z_i; i = 1, 2, \ldots\}$ is an i.i.d. sequence with $Z_i =^d U_\delta$. We stop this procedure upon finding the integer N satisfying

$$T \in [U_{N\delta}, U_{(N+1)\delta}). \tag{6.53}$$

Note that the $\mathbb{N} \cup \{0\}$-valued random variable N indeed exists since $U_t \to \infty$ as $t \to \infty$ with probability one. To generate the random variables $\{Z_i\}$, one can use various algorithms presented in [Cont and Tankov (2003), Chap. 6], which include those based on series representations of Lévy processes discussed in [Rosiński (2001)]. For simulation of tempered stable random variables, one can also consult [Bauemer and Meerschaert (2010)]. Next, let

$$E_t^\delta := (\min\{n \in \mathbb{N}; U_{n\delta} > t\} - 1)\delta, \quad t \in [0, T]. \tag{6.54}$$

The sample paths of E^δ are nondecreasing step functions with constant jump size δ and the ith waiting time given by $Z_i = U_{i\delta} - U_{(i-1)\delta}$. Indeed, it is easy to see that for $n = 0, 1, 2, \ldots, N$,

$$E_t^\delta = n\delta \text{ whenever } t \in [U_{n\delta}, U_{(n+1)\delta}). \tag{6.55}$$

In particular, (6.53) is equivalent to

$$E_T^\delta = N\delta. \tag{6.56}$$

As mentioned earlier, to approximate the solution $(X_t)_{t \in [0, T]}$ of SDE (6.46), which can be written as $X_t = Y_{E_t}$, we will use the process $(X_t^\delta)_{t \in [0, T]}$ defined by

$$X_t^\delta := Y_{E_t^\delta}^\delta.$$

A natural question to ask is whether X^δ converges to X in some reasonable sense as $\delta \to 0$ and, if so, what the rate of convergence is. To state theorems which answer the question, let us introduce various notions of order of convergence (see [Kloeden and Platen (1999)] for details):

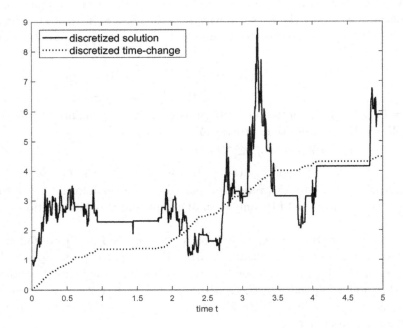

Fig. 6.1   Simulation of a sample path of the solution of a time-changed Black–Scholes SDE $dX_t = X_t\,dE_t + X_t\,dB_{E_t}$ with initial condition $X_0 = 1$ on the time interval $[0,5]$, along with a sample path of an inverse stable subordinator $E_t = E_t^{\beta}$ with index $\beta = 0.85$. The time step is taken to be $\delta = 10^{-3}$.

- X^{δ} is said to *converge strongly to X at time T with order* $\gamma \in (0, \infty)$ if there exist finite positive constants C and δ_0 such that for all $\delta \in (0, \delta_0)$,

$$\mathbb{E}[|X_T - X_T^{\delta}|] \leqslant C\delta^{\gamma}.$$

- X^{δ} is said to *converge strongly to X uniformly on $[0, T]$ with order* $\gamma \in (0, \infty)$ if there exist finite positive constants C and δ_0 such that for all $\delta \in (0, \delta_0)$,

$$\mathbb{E}\left[\sup_{0 \leqslant t \leqslant T} |X_t - X_t^{\delta}|\right] \leqslant C\delta^{\gamma}.$$

- X^{δ} is said to *converge weakly to X at time T with order* $\gamma \in (0, \infty)$ if for any function g in a suitable function space, there exist finite positive constants C and δ_0 such that for all $\delta \in (0, \delta_0)$,

$$\left|\mathbb{E}[g(X_T) - g(X_T^{\delta})]\right| \leqslant C\delta^{\gamma}.$$

Theorem 6.14 (Strong convergence). *There exists a finite positive constant C such that for all $\delta \in (0,1)$,*

$$\mathbb{E}[|X_T - X_T^\delta|^2] \leqslant C\delta,$$

which in particular implies $\mathbb{E}[|X_T - X_T^\delta|] \leqslant C^{1/2}\delta^{1/2}$. Thus, X^δ converges strongly to X at the time horizon T with order $1/2$.

Theorem 6.15 (Uniform strong convergence). *Let $\varepsilon \in (0,1)$. There exist finite positive constants C and $\delta_0 \in (0,1)$ such that for all $\delta \in (0,\delta_0)$,*

$$\mathbb{E}\left[\sup_{0 \leqslant t \leqslant T} |X_t - X_t^\delta|^2\right] \leqslant C\delta^{1-\varepsilon},$$

which in particular implies $\mathbb{E}\left[\sup_{0 \leqslant t \leqslant T} |X_t - X_t^\delta|\right] \leqslant C^{1/2}\delta^{(1-\varepsilon)/2}$. Thus, X^δ converges strongly to X uniformly on $[0,T]$ with any order $\gamma \in (0,1/2)$.

Theorem 6.16 (Weak convergence). *Suppose that SDE (6.46) on $[0,T]$ has autonomous coefficients $b(t,x) = b(x)$ and $\sigma(t,x) = \sigma(x)$ satisfying conditions (6.47) and (6.48). Assume further that the coefficients are in $C^4(\mathbb{R}^d)$ and have derivatives of polynomial growth. Let $g \in C^4(\mathbb{R}^d)$ have derivatives of polynomial growth. Then there exists a finite positive constant C such that for all $\delta \in (0,1)$,*

$$\left|\mathbb{E}[g(X_T)] - g(X_T^\delta)]\right| \leqslant C\delta. \tag{6.57}$$

Thus, X^δ converges weakly to X at the time horizon T with order 1.

For the proofs of these results, consult [Jum and Kobayashi (2016), Thm. 3.1-3.3]. Note in particular that Theorem 6.15 is an immediate consequence of the proof of [Jum and Kobayashi (2016), Thm. 3.2]. Analysis of errors involved with a specific numerical example is provided in Section 4 of that paper.

Chapter 7

Fractional Fokker–Planck–Kolmogorov equations

Introduction

This chapter describes various classes of stochastic processes defined by stochastic differential equations (SDEs) whose associated Fokker–Planck–Kolmogorov (FPK) equations are time-fractional order partial or pseudo-differential equations. These SDEs are driven by time-changed processes, where the time-change is the inverse to a stable subordinator or to a mixture of independent stable subordinators. In Section 7.1, we first recall basic results on the classical Itô diffusion. Section 7.2 discusses FPK equations associated with SDEs driven by a mixture of time-changed Lévy processes. These FPK equations are time-fractional order pseudo-differential equations. Section 7.3 provides necessary preliminaries on FPK equations associated with fractional Brownian motion. We also present an operator approach for derivation of time-fractional FPK equations associated with SDEs driven by a time-changed Brownian motion. In Section 7.4, this approach is extended to the case of a time-changed fractional Brownian motion. Sections 7.5–7.6 deal with more general Gaussian and time-changed Gaussian processes. In Section 7.7, we derive time-fractional FPK equations associated with stochastic processes in bounded domains. Sections 7.2, 7.4 and 7.6 each contain important applications that can be deduced from the prior material.

Figure 1.1 of Chapter 1 presents a general paradigm with a driving process [DP] in the middle surrounded by a triangle whose vertices are [FPK], [CTRW] and [SDE]. The material given in Chapter 6 together with the discussions to be given in this chapter should clarify interconnections between those vertices as well as their connections with a driving process or family of driving processes.

In Sections 7.1 and 7.2, we will use the following shorthand notations to describe different kinds of FPK equations:

- *FPKΨ*: an FPK equation with a pseudo-differential operator
- *TFFPK*: a time-fractional FPK equation
- *TDFPK*: a time-fractional distributed order FPK equation
- *TFFPKΨ*: a time-fractional FPK equation with a pseudo-differential operator
- *TDFPKΨ*: a time-fractional distributed order FPK equation with a pseudo-

differential operator.

7.1 FPK and FPKΨ equations associated with SDEs driven by Brownian motion and Lévy processes

Let $(\Omega, \mathcal{F}, \mathbb{P})$ be a probability space with a complete right-continuous filtration $(\mathcal{F}_t)$ and $Z_t = B_t + bt$ be an n-dimensional Brownian motion with constant drift bt, $b \in \mathbb{R}^n$, defined on this filtered probability space. Let $P^Z(t, x, \Gamma) = \mathbb{P}(Z_t \in \Gamma | Z_0 = x)$, $\Gamma \in \mathcal{B}(\mathbb{R}^n)$, be the transition probability of the process Z_t with density $p^Z(t, x, y)$, i.e. $p^Z(t, x, y)\, dy = P^Z(t, x, dy)$. Then $p^Z(t, x, y)$ satisfies in the weak sense (in the sense of distributions) the following partial differential equation (see, e.g. [Stroock (2003)]):

$$\frac{\partial p^Z(t, x, y)}{\partial t} = -\sum_{j=1}^{n} b_j \frac{\partial p^Z(t, x, y)}{\partial y_j} + \frac{1}{2} \sum_{j=1}^{n} \frac{\partial^2 p^Z(t, x, y)}{\partial y_j^2}, \quad t > 0, \ x, y \in \mathbb{R}^n,$$

with the additional condition $p^Z(0, x, y) = \delta_x(y)$, where δ_x is the Dirac delta function with mass on x. A deep generalization of this relationship between a stochastic process and its associated partial differential equation was expressed through the Kolmogorov forward and backward equations. This concept is based on the relationship between two main components:

(i) the Cauchy problem

$$\frac{\partial u(t, x)}{\partial t} = \mathcal{A}u(t, x), \ t > 0, \ x \in \mathbb{R}^n, \tag{7.1}$$

$$u(0, x) = \varphi(x), \ x \in \mathbb{R}^n; \tag{7.2}$$

where $\mathcal{A}$ is a differential operator

$$\mathcal{A} = \sum_{j=1}^{n} b_j(x) \frac{\partial}{\partial x_j} + \frac{1}{2} \sum_{i,j=1}^{n} \sigma_{i,j}(x) \frac{\partial^2}{\partial x_i \partial x_j}, \tag{7.3}$$

with coefficients $b_j(x)$ and $\sigma_{i,j}(x)$ satisfying some mild regularity conditions; and

(ii) the associated class of Itô SDEs given by

$$dX_t = b(X_t)dt + \sigma(X_t)dB_t, \ X_0 = x, \tag{7.4}$$

where B_t is an m-dimensional Brownian motion. Here X_t is a solution, and the coefficients are connected with the coefficients of the operator $\mathcal{A}$ as follows: $b(x) = (b_1(x), \ldots, b_n(x))$ and $\sigma_{i,j}(x)$ is the (i, j)-th entry of the product of the $n \times m$ matrix $\sigma(x)$ with its transpose $\sigma^T(x)$.

One mechanism for establishing this relationship is via semigroup theory, in which the operator $\mathcal{A}$ is recognized as the infinitesimal generator of the semigroup $T_t(\cdot)(x) := \mathbb{E}[(\cdot)(X_t)|X_0 = x]$ (defined, for instance, on the Banach space $C_0(\mathbb{R}^n)$ with sup-norm), i.e. $\mathcal{A}\varphi(x) = \lim_{t \to 0} (T_t - I)\varphi(x)/t$, $\varphi \in \mathrm{Dom}(\mathcal{A})$, the domain of $\mathcal{A}$. A unique solution to (7.1)–(7.2) for $\mathcal{A}$ in (7.3) is represented by $u(t, x) = (T_t\varphi)(x)$.

The relationship between (i) and (ii) says that the class of deterministic partial differential equations given in (7.1), with a first order time derivative on the left and the operator $\mathcal{A}$ specified in (7.3) on the right, is associated to the class of SDEs in (7.4) driven by a Brownian motion with drift as long as the coefficients satisfy appropriate conditions. The mechanism for establishing the relationship reveals that the transition probabilities $P^X(t, x, dy) = \mathbb{P}(X_t \in dy | X_0 = x)$ of a solution X_t to (7.4) satisfy in the weak sense the following partial differential equations (see, e.g. [Bell (1995)]):

$$\frac{\partial P^X(t, x, dy)}{\partial t} = \mathcal{A}\, P^X(t, x, dy), \qquad (\mathcal{A} \text{ acts on the variable } x) \qquad (7.5)$$

$$\frac{\partial P^X(t, x, dy)}{\partial t} = \mathcal{A}^*\, P^X(t, x, dy), \qquad (\mathcal{A}^* \text{ acts on the variable } y) \qquad (7.6)$$

where $\mathcal{A}^*$ is the formal conjugate to $\mathcal{A}$. Equation (7.5), in which $\mathcal{A}$ acts on the backward variable x, is called a backward Kolmogorov equation. Equation (7.6), where $\mathcal{A}^*$ acts on the forward variable y, is called a forward Kolmogorov equation or, in the physics literature, a Fokker–Planck equation. The two equations are referred to as Fokker–Planck–Kolmogorov (FPK) equations.

The relationship between the stochastic process X_t in (7.4) and another associated partial differential equation

$$\frac{\partial w}{\partial t} = \mathcal{A}w - qw, \quad w(0, x) = \varphi(x), \qquad (7.7)$$

for a nonnegative continuous function q is given by the Feynman–Kac formula:

$$w(t, x) = \mathbb{E}\left[\exp\left(-\int_0^t q(X_s)ds \right) \varphi(X_t) \Big| X_0 = x \right]. \qquad (7.8)$$

Enlarging the SDEs in (7.4) to those driven by a Lévy process leads to a generalization of relationship (i)–(ii) where the analogous operator on the right hand side of (7.1) has additional terms corresponding to jump components of the driving process (see [Applebaum (2009), Situ (2005)] and references therein). In this case, the operator $\mathcal{A}$ in (7.3) takes the form

$$\mathcal{A}\varphi(x) = \mathcal{A}(x, D_x)\varphi(x)$$

$$:= \sum_{j=1}^n b_j(x) \frac{\partial \varphi(x)}{\partial x_j} + \frac{1}{2} \sum_{i,j=1}^n \sigma_{i,j}(x) \frac{\partial^2 \varphi(x)}{\partial x_i \partial x_j}$$

$$+ \int_{\mathbb{R}^n \setminus \{0\}} \left[\varphi(x + G(x, w)) - \varphi(x) - \boldsymbol{I}_{(|w|<1)}(w) \sum_{j=1}^n G_j(x, w) \frac{\partial \varphi(x)}{\partial x_j} \right] \nu(dw), \quad (7.9)$$

where ν is a Lévy measure, $G(x, w)$ is a given vector-function (see Section 5.3 for details), and $\boldsymbol{I}_B$ stands for the indicator function of a set B. The operator $\mathcal{A}(x, D_x)$ can be considered as a pseudo-differential operator with a symbol $\Psi(x, \xi)$ (given in (5.21)) defined in accordance with a Brownian motion with drift and a Lévy measure that controls jump components of the underlying Lévy process. The corresponding FPK equation

$$\frac{\partial u(t,x)}{\partial t} = \mathcal{A}(x, D_x)u(t,x), \ t > 0, \ x \in \mathbb{R}^n, \tag{7.10}$$

is referred to as an *FPKΨ* equation with "Ψ" to emphasize the appearance of the pseudo-differential operator on the right.

7.2 TFFPKΨ/TDFPKΨ equations associated with SDEs driven by time-changed Lévy processes

In this section we generalize the relationship (i)–(ii) in Section 7.1 to *TFFPKΨ/TDFPKΨ* equations that imply time-fractional analogues of the *FPKΨ* equation (7.10). First, the generalization establishes the class of SDEs replacing (7.4) which is associated with the following Cauchy problem:

$$D_*^\beta u(t,x) = \mathcal{A}(x, D_x)u(t,x), \ t > 0, \ x \in \mathbb{R}^n, \tag{7.11}$$

$$u(0,x) = \varphi(x), \ x \in \mathbb{R}^n, \tag{7.12}$$

in which the classes of allowable operators on the left and on the right of (7.11) are respectively D_*^β, a fractional derivative in the sense of Caputo–Djrbashian with $\beta \in (0,1)$ (see Chapter 4) and $\mathcal{A}(x, D_x)$, a pseudo-differential operator in (7.9) with Lévy symbol $\Psi(x, \xi)$ in (5.21). Unlike (7.1) and (7.10), equation (7.11) has the structure of *TFFPKΨ* equations.

The driving processes of the associated class of SDEs are Lévy processes composed with the inverse of a β-stable subordinator, $\beta \in (0,1)$ (Theorem 7.3 with $N = 1$). Since such processes are semimartingales, SDEs with respect to them are meaningful and have the form

$$X_t = X_0 + \int_0^t b(X_{s-})dE_s + \int_0^t \sigma(X_{s-})dB_{E_s}$$

$$+ \int_0^t \int_{\mathbb{R}^n \setminus \{0\}} G(X_{s-}, w)N(dE_s, dw), \ t > 0,$$

where N is a Poisson random measure, E_t is the the first hitting time for a stable subordinator and the functions $b(x)$, $\sigma(x)$ and $G(x,w)$ satisfy conditions to be specified later. This completes the relationship (analogue of (i)–(ii)) which includes the Cauchy problem in (7.11)–(7.12) given through a *TFFPKΨ* equation. A partial result when the driving process is either a Brownian or stable Lévy motion with drift time-changed by the inverse of a stable subordinator is considered in [Magdziarz and Weron (2007), Magdziarz et al. (2008)] without specifying the explicit form of the corresponding SDEs.

More generally, the class of SDEs in the above discussion when E_t is the inverse of an arbitrary mixture of independent stable subordinators gives rise to a Cauchy problem with a time-fractional derivative with distributed orders on the left of (7.11), namely, the *TDFPKΨ* equation

$$D_\mu u(t,x) = \mathcal{A}(x, D_x)u(t,x).$$

In this case, the time-change E_t is no longer the inverse of a stable subordinator if at least two different indices arise in the mixture. Moreover, SDEs corresponding to time-fractional FPK equations cannot be described within the classical Brownian- or Lévy-driven SDEs. From this point of view, the theory of stochastic integrals and stochastic differential equations driven by time-changed semimartingales in Sections 6.1 and 6.2 plays an important role in the following discussion.

7.2.1 *Theory*

In order to establish *TFFPKΨ/TDFPKΨ* equations, first we prove two abstract theorems. Theorems 7.1 and 7.2 require the following *assumption*: $\{T_t, t \geq 0\}$ is a strongly continuous semigroup defined on a Banach space $\mathcal{X}$ with norm $\| \cdot \|$ such that the estimate

$$\|T_t \varphi\| \leq M \|\varphi\| e^{\omega t} \tag{7.13}$$

is valid for some constants $M \geq 1$ and $\omega \geq 0$.

This assumption implies that any number s with $Re(s) > \omega$ belongs to the resolvent set $\rho(\mathcal{A})$ of the infinitesimal generator $\mathcal{A}$ of T_t and the resolvent operator is represented in the form $R(s, \mathcal{A}) = \int_0^\infty e^{-st} T_t \, dt$ (see e.g. [Engel and Nagel (1999)]).

Theorem 7.1. *Let* $U_t = \sum_{k=1}^{N} c_k U_{k,t}$, *where* $U_{k,t}$, $k = 1, \ldots, N$, *are independent stable subordinators with respective indices* $\beta_k \in (0,1)$ *and constants* $c_k > 0$. *Let* E_t *be the inverse process to* U_t. *Suppose* $\{T_t, t \geq 0\}$ *is a strongly continuous semigroup in a Banach space* $\mathcal{X}$, *satisfies* (7.13), *and has infinitesimal generator* $\mathcal{A}$ *with* $Dom(\mathcal{A}) \subset \mathcal{X}$. *Then for each fixed* $t \geq 0$ *and* $\varphi \in Dom(\mathcal{A})$, *the integral* $\int_0^\infty f_{E_t}(\tau) T_\tau \varphi \, d\tau$ *exists and the vector-function* $v(t) = \int_0^\infty f_{E_t}(\tau) T_\tau \varphi \, d\tau$ *satisfies the abstract Cauchy problem*

$$\sum_{k=1}^{N} C_k D_*^{\beta_k} v(t) = \mathcal{A}v(t), \ t > 0, \tag{7.14}$$

$$v(0) = \varphi, \tag{7.15}$$

where $C_k = c_k^{\beta_k}$, $k = 1, \ldots, N$.

Proof. For simplicity, the proof will be given in the case $N = 2$. First, define a vector-function $p(\tau) = T_\tau \varphi$, where $\varphi \in Dom(\mathcal{A})$. In accordance with the conditions of the theorem, $p(\tau)$ satisfies the abstract Cauchy problem

$$\frac{\partial p(\tau)}{\partial \tau} = \mathcal{A}p(\tau), \ p(0) = \varphi, \tag{7.16}$$

where the operator $\mathcal{A}$ is the infinitesimal generator of T_τ. Now consider the integral $\int_0^\infty f_{E_t}(\tau) T_\tau \varphi \, d\tau$. It follows from Lemma 5.4 and condition (7.13) that

$$\left\| \int_0^\infty f_{E_t}(\tau) T_\tau \varphi \, d\tau \right\| \leq \int_0^\infty f_{E_t}(\tau) \|T_\tau \varphi\| \, d\tau \tag{7.17}$$

$$\leq C \|\varphi\| \int_0^\infty e^{-(K\tau^{\frac{1}{1-\beta}} - \omega\tau)} \, d\tau < \infty,$$

where $\beta \in (0, 1)$ and C, $K > 0$ are constants. Hence, the integral $\int_0^\infty f_{E_t}(\tau) T_\tau \varphi \, d\tau$ exists in the sense of Bochner for each fixed $t \geq 0$. Denote this vector-function by

$$v(t) := \int_0^\infty f_{E_t}(\tau) T_\tau \varphi \, d\tau. \tag{7.18}$$

It follows immediately from the definition of the semigroup T_t that

$$v(0) = \lim_{t \to 0+} \int_0^\infty f_{E_t}(\tau) T_\tau \varphi \, d\tau = T_0 \varphi = \varphi,$$

in the norm of $\mathcal{X}$. By (5.40),

$$v(t) = -\int_0^\infty \frac{\partial}{\partial \tau} \left\{ \frac{1}{c_2 \tau^{\frac{1}{\beta_2}}} \left[(Jf_1^{(1)}) \left(\frac{\cdot}{c_1 \tau^{\frac{1}{\beta_1}}} \right) * f_1^{(2)} \left(\frac{\cdot}{c_2 \tau^{\frac{1}{\beta_2}}} \right) \right](t) \right\} T_\tau \varphi \, d\tau.$$

To calculate the Laplace transform $L[v(t)](s) = \tilde{v}(s) = \int_0^\infty e^{-st} v(t) \, dt$, note that

$$L\left[\frac{1}{b} (Jf_1^{(1)}) \left(\frac{t}{a} \right) * f_1^{(2)} \left(\frac{t}{b} \right) \right](s) = \frac{1}{b} \frac{1}{as} \left(a \widetilde{f_1^{(1)}}(as) \right) \left(b \widetilde{f_1^{(2)}}(bs) \right)$$

$$= \frac{1}{s} \widetilde{f_1^{(1)}}(as) \widetilde{f_1^{(2)}}(bs).$$

Then it follows from (5.26) that

$$\tilde{v}(s) = -\int_0^\infty \frac{\partial}{\partial \tau} \left\{ \frac{1}{s} e^{-\tau c_1^{\beta_1} s^{\beta_1}} e^{-\tau c_2^{\beta_2} s^{\beta_2}} \right\} T_\tau \varphi \, d\tau \tag{7.19}$$

$$= (c_1^{\beta_1} s^{\beta_1 - 1} + c_2^{\beta_2} s^{\beta_2 - 1}) \int_0^\infty e^{-\tau (c_1^{\beta_1} s^{\beta_1} + c_2^{\beta_2} s^{\beta_2})} T_\tau \varphi \, d\tau$$

$$= (C_1 s^{\beta_1 - 1} + C_2 s^{\beta_2 - 1}) \widetilde{[T_\tau \varphi]} (C_1 s^{\beta_1} + C_2 s^{\beta_2})$$

$$= (C_1 s^{\beta_1 - 1} + C_2 s^{\beta_2 - 1}) \tilde{p} (C_1 s^{\beta_1} + C_2 s^{\beta_2}),$$

which is well defined for all s such that $C_1 s^{\beta_1} + C_2 s^{\beta_2} > \omega \geq 0$, where $C_k = c_k^{\beta_k}$, $k = 1, 2$. Let $\omega_0 \geq 0$ be a number such that $s > \omega_0$ if and only if $C_1 s^{\beta_1} + C_2 s^{\beta_2} > \omega$. Equation (7.19) implies

$$\tilde{p}(C_1 s^{\beta_1} + C_2 s^{\beta_2}) = \frac{\tilde{v}(s)}{C_1 s^{\beta_1 - 1} + C_2 s^{\beta_2 - 1}}, \quad s > \omega_0. \tag{7.20}$$

On the other hand it follows from (7.16) that

$$(s - A)\tilde{p}(s) = \varphi, \quad s > \omega. \tag{7.21}$$

Then equations (7.19), (7.20), and (7.21) together yield

$$[C_1 s^{\beta_1} + C_2 s^{\beta_2} - A]\tilde{v}(s) = (C_1 s^{\beta_1 - 1} + C_2 s^{\beta_2 - 1})\varphi, \quad s > \omega_0,$$

which can be rewritten as

$$C_1[s^{\beta_1} \tilde{v}(s) - s^{\beta_1 - 1} v(0)] + C_2[s^{\beta_2} \tilde{v}(s) - s^{\beta_2 - 1} v(0)] = A\tilde{v}(s), \quad s > \omega_0. \tag{7.22}$$

Applying the inverse Laplace transform L^{-1} to both sides of (7.22) and recalling the fact (see (3.22))

$$L^{-1}[s^\alpha \tilde{v}(s) - s^{\alpha - 1} v(0)](t) = D_*^\alpha v(t), \quad 0 < \alpha < 1,$$

yields

$$C_1 D_*^{\beta_1} v(t) + C_2 D_*^{\beta_2} v(t) = Av(t).$$

Hence $v(t)$ satisfies the Cauchy problem (7.14)–(7.15). The case $N > 2$ can be proved using the same method. $\qquad \square$

The next theorem provides an extension with U_t as the weighted average of an arbitrary number of independent stable subordinators. It is easy to verify that the process $U_t = \sum_{k=1}^{N} c_k U_{k,t}$ given in Theorem 7.1 satisfies

$$\ln \mathbb{E}\big[e^{-sU_t}\big]\Big|_{t=1} = -\sum_{k=1}^{N} c_k^{\beta_k} s^{\beta_k}, \quad s \geqslant 0. \tag{7.23}$$

The function on the right hand side of (7.23) can be expressed as the integral $-\int_0^1 s^\beta d\mu(\beta)$, with μ the finite atomic measure, $d\mu(\beta) = \sum_{k=1}^{N} c_k^{\beta_k} \delta_{\beta_k}(\beta) d\beta$. The integral $\int_0^1 s^\beta d\mu(\beta)$ is meaningful for any finite measure μ defined on $(0,1)$. Let $\mathcal{S}$ designate the class of strictly increasing processes V_t whose Laplace transform is given by

$$\ln \mathbb{E}\big[e^{-sV_t}\big] = -t \int_0^1 s^\beta d\mu(\beta), \quad s \geqslant 0, \tag{7.24}$$

where μ is a finite positive measure defined on the interval $(0,1)$. This class of stochastic processes obviously contains stable subordinators and all mixtures of finitely many independent stable subordinators. Note that $V_0 = 0$ a.s. by construction. For the process $V_t \in \mathcal{S}$ corresponding to a finite measure μ, we use the notation $V_t = U(\mu,t)$ to indicate this correspondence. The inverse of $U(\mu,t) \in \mathcal{S}$ is denoted as $E(\mu,t)$.

One can establish a relationship between $U(\mu,t)$ and U_t^β, $0 < \beta < 1$. Namely, it follows from the Laplace transforms (5.26) and (7.24) that the stochastic processes $U(\mu,t)$ and U_t^β are related through the relationship

$$\ln \mathbb{E}\big[e^{-sU(\mu,t)}\big] = \int_0^1 \ln\big[e^{-sU_t^\beta}\big] d\mu(\beta).$$

The latter implies that the corresponding density functions $f_{U(\mu,t)}(\tau)$ and $f_{U_t^\beta}(\tau)$ are related through the equation

$$\int_0^\infty f_{U(\mu,t)}(\tau)e^{-s\tau}\, d\tau = e^{\int_0^1 \ln\left(\int_0^\infty f_{U_t^\beta}(\tau)e^{-s\tau}d\tau\right) d\mu(\beta)},$$

or equivalently,

$$f_{U(\mu,t)}(\tau) = L_{s\to\tau}^{-1}\left[e^{\int_0^1 \ln\left(\int_0^\infty f_{U_t^\beta}(\tau)e^{-s\tau}d\tau\right) d\mu(\beta)}\right](\tau),$$

where $L_{s\to\tau}^{-1}$ stands for the inverse Laplace transform with the resulting function of τ for each fixed $t \geqslant 0$.

Remark 7.1. A direct relationship between the inverses $E(\mu,t)$ and E_t^β, $0 < \beta < 1$, is currently unknown.

Theorem 7.2. *Let $E(\mu,t)$ be the inverse of a subordinator $U(\mu,t) \in \mathcal{S}$ and let $\{T_t\}$ and φ be as in Theorem 7.1. Then the vector-function $v(t) = \int_0^\infty f_{E(\mu,t)}(\tau)T_\tau\varphi\,d\tau$ exists and satisfies the abstract Cauchy problem*

$$D_\mu v(t) := \int_0^1 D_*^\beta v(t)d\mu(\beta) = Av(t), \quad t > 0, \tag{7.25}$$

$$v(0) = \varphi. \tag{7.26}$$

Proof. We briefly sketch the proof, since the idea is similar to the proof of Theorem 7.1. The density $f_{U(\mu,t)}(\tau), \tau \geq 0$, exists and has asymptotics (5.27) with some $\beta = \beta_0 \in (0,1)$ and (5.28) with some $\beta = \beta_1 \in (0,1)$. This implies existence of the vector-function $v(t)$. Further, one can readily see that

$$v(t) = -\int_0^\infty \frac{\partial}{\partial \tau}\{J f_{U(\mu,\tau)}(t)\}(T_\tau \varphi)d\tau.$$

Now it follows from the definition of $U(\mu,t)$ that the Laplace transform of $v(t)$ satisfies

$$\tilde{v}(s) = \frac{\int_0^1 s^\beta d\mu(\beta)}{s}\int_0^\infty e^{-\tau \int_0^1 s^\beta d\mu(\beta)}(T_\tau \varphi)d\tau = \frac{\Phi_\mu(s)}{s}\tilde{p}(\Phi_\mu(s)), \quad s > \bar{\omega}, \qquad (7.27)$$

where

$$\Phi_\mu(s) = \int_0^1 s^\beta d\mu(\beta), \qquad (7.28)$$

and p is a solution to the abstract Cauchy problem (7.16). Here $\bar{\omega} > 0$ is a number such that $s > \bar{\omega}$ if $\Phi_\mu(s) > \omega$ ($\bar{\omega}$ is uniquely defined, since as is seen from (7.28), the function $\Phi_\mu(s)$ is a strictly increasing function). We have seen (see (7.21)) that $\tilde{p}(s)$ satisfies the equation $(s - \mathcal{A})\tilde{p}(s) = \varphi$ for all $s > \omega$, whereas (7.27) implies that $\tilde{p}(\Phi_\mu(s)) = s\tilde{v}(s)/\Phi_\mu(s)$ for all $s > \bar{\omega}$. Therefore,

$$(\Phi_\mu(s) - \mathcal{A})\frac{s\tilde{v}(s)}{\Phi_\mu(s)} = \varphi, \quad s > \bar{\omega},$$

or

$$(\Phi_\mu(s) - \mathcal{A})\tilde{v}(s) = \varphi\frac{\Phi_\mu(s)}{s}, \quad s > \bar{\omega}.$$

We rewrite the latter in the form

$$\Phi_\mu(s)\tilde{v}(s) - \frac{\Phi_\mu(s)}{s}v(0) = \mathcal{A}\tilde{v}(s), \quad s > \bar{\omega}. \qquad (7.29)$$

Due to formula (3.31), taking the inverse Laplace transform to both sides of (7.29) yields (7.25), as desired.

Finally, using the fact that $E(\mu, 0) = 0$ a.s. and the dominated convergence theorem yields

$$\lim_{t \to 0} v(t) = \int_0^\infty \delta_0(\tau)T_\tau \varphi d\tau = T_0 \varphi = \varphi,$$

which completes the proof. □

Now we consider fractional FPK equations associated with SDEs driven by a time-changed Lévy process. For SDEs driven by a Lévy process, see Section 5.3 for details.

Theorem 7.3. *Let* $U_{k,t}, k = 1, \ldots, N$ *be independent stable subordinators of respective indices* $\beta_k \in (0,1)$. *Define* $U_t = \sum_{k=1}^N c_k U_{k,t}$, *with positive constants* c_k, *and let E_t be its inverse. Suppose that a stochastic process Y_τ satisfies the SDE driven by a Lévy process*

$$Y_\tau = x + \int_0^\tau b(Y_{s-})ds + \int_0^\tau \sigma(Y_{s-})dB_s \tag{7.30}$$

$$+ \int_0^\tau \int_{|w|<1} H(Y_{s-}, w)\tilde{N}(ds, dw) + \int_0^\tau \int_{|w|\geqslant 1} K(Y_{s-}, w)N(ds, dw),$$

where the continuous mappings $b : \mathbb{R}^n \to \mathbb{R}^n$, $\sigma : \mathbb{R}^n \to \mathbb{R}^{n\times m}$, $H : \mathbb{R}^n \times \mathbb{R}^n \to \mathbb{R}^n$, and $K : \mathbb{R}^n \times \mathbb{R}^n \to \mathbb{R}^n$ are bounded and satisfy condition (5.18). Let $X_t = Y_{E_t}$. Then

1) *X_t satisfies the SDE driven by the time-changed Lévy process*

$$X_t = x + \int_0^t b(X_{s-})dE_s + \int_0^t \sigma(X_{s-})dB_{E_s} \tag{7.31}$$

$$+ \int_0^t \int_{|w|<1} H(X_{s-}, w)\tilde{N}(dE_s, dw) + \int_0^t \int_{|w|\geqslant 1} K(X_{s-}, w)N(dE_s, dw);$$

2) *if the process Y_τ is independent of the process E_t, then the function $u(t, x) = \mathbb{E}[\varphi(X_t)|X_0 = x]$, where $\varphi \in C_0^2(\mathbb{R}^n)$, satisfies the following Cauchy problem for the TDFPKΨ equation*

$$\sum_{k=1}^N C_k D_*^{\beta_k} u(t, x) = \mathcal{A}(x, D_x)u(t, x), \ t > 0, \ x \in \mathbb{R}^n, \tag{7.32}$$

$$u(0, x) = \varphi(x), \ x \in \mathbb{R}^n. \tag{7.33}$$

Here $C_k = c_k^{\beta_k}$, $k = 1, \ldots, N$, and $\mathcal{A}(x, D_x)$ is the pseudo-differential operator

$$\mathcal{A}(x, D_x)\varphi(x) = i(b(x), D_x)\varphi(x) - \frac{1}{2}(\Sigma(x)D_x, D_x)\varphi(x) \tag{7.34}$$

$$+ \int_{\mathbb{R}^n\setminus\{0\}} [\varphi(x + G(x, w)) - \varphi(x) - iI_{(-1,1)}(w)(G(x, w), D_x)\varphi(x)]\nu(dw)$$

with symbol

$$\Psi(x, \xi) = i(b(x), \xi) - \frac{1}{2}(\Sigma(x)\xi, \xi) \tag{7.35}$$

$$+ \int_{\mathbb{R}^n\setminus\{0\}} (e^{i(G(x,w),\xi)} - 1 - i(G(x, w), \xi)I_{(-1,1)}(w))\nu(dw),$$

where $\Sigma(x) = \sigma(x)\times\sigma(x)^T$ and $G(x, w) = H(x, w)$ if $|w| < 1$ and $G(x, w) = K(x, w)$ if $|w| \geqslant 1$.

Proof. Again, for simplicity, we give the proof in the case $N = 2$. The proof of part 1) easily follows from the Duality Theorem (Theorem 6.2). Notice that since U_t is a linear combination of stable subordinators, which are càdlàg and strictly increasing, U_t is also càdlàg and strictly increasing. Hence, $X_t = Y_{E_t}$ satisfies SDE (7.31).

2) Consider $T_\tau^Y \varphi(x) = \mathbb{E}[\varphi(Y_\tau)|Y_0 = x]$, where Y_τ is a solution of SDE (7.30). Then T_τ^Y is a strongly continuous contraction semigroup in the Banach space $C_0(\mathbb{R}^n)$ (see [Applebaum (2009)]) which satisfies (7.13) with $\omega = 0$, has infinitesimal generator given by the pseudo-differential operator $\mathcal{A}(x, D_x)$ with symbol $\Psi(x, \xi)$ defined in (7.35), and $C_0^2(\mathbb{R}^n) \subset \text{Dom}(\mathcal{A}(x, D_x))$. So the function $p^Y(\tau, x) = T_\tau^Y \varphi(x)$ with $\varphi \in C_0^2(\mathbb{R}^n)$ satisfies the Cauchy problem

$$\frac{\partial p^Y(\tau, x)}{\partial \tau} = \mathcal{A}(x, D_x)p^Y(\tau, x), \quad p^Y(0, x) = \varphi(x). \tag{7.36}$$

Furthermore, consider $p^X(t, x) = \mathbb{E}[\varphi(X_t)|X_0 = x] = \mathbb{E}[\varphi(Y_{E_t})|Y_0 = x]$ (recall that $E_0 = 0$). Using independence of the processes Y_τ and E_t,

$$p^X(t, x) = \int_0^\infty \mathbb{E}[\varphi(Y_\tau)|E_t = \tau, Y_0 = x] f_{E_t}(\tau) d\tau = \int_0^\infty f_{E_t}(\tau) T_\tau^Y \varphi(x) d\tau. \tag{7.37}$$

Now, in accordance with Theorem 7.1, $p^X(t, x)$ satisfies the Cauchy problem (7.32)–(7.33). $\qquad\square$

Theorem 7.4. *Let $E(\mu, t)$ be the inverse of a subordinator $U(\mu, t) \in \mathcal{S}$. Suppose that a stochastic process Y_τ satisfies SDE (7.30), and let $X_t = Y_{E(\mu,t)}$. Then*

1) *X_t satisfies SDE (7.31);*
2) *if Y_τ is independent of $E(\mu, t)$, then the function $u(t, x) = \mathbb{E}[\varphi(X_t)|X_0 = x]$ satisfies the following Cauchy problem for the TDFPKΨ equation*

$$D_\mu u(t, x) = \mathcal{A}(x, D_x)u(t, x), \ t > 0, \ x \in \mathbb{R}^n, \tag{7.38}$$

$$u(0, x) = \varphi(x), \ x \in \mathbb{R}^n. \tag{7.39}$$

Proof. The proof of part 1) again follows from Theorem 6.2. Part 2) follows from Theorem 7.2 in a manner similar to the proof of part 2) of Theorem 7.3. $\qquad\square$

Remark 7.2. Theorems 7.3 and 7.4 reveal the class of SDEs which are associated with the wide class of *TDFPKΨ* equations. Each SDE in this class is driven by a semimartingale which is a time-changed Lévy process, where the time-change is given by the inverse of a mixture of independent stable subordinators. Therefore, these SDEs cannot be represented as classical SDEs driven by a Brownian motion or a Lévy process. The general extensions provided by these two theorems were motivated by their requirement in many applications, such as the cell biology example considered in Chapter 1.

Corollary 7.5. *Let the coefficients b, σ, H, K of the pseudo-differential operator $\mathcal{A}(x, D_x)$ defined in (7.34) with symbol in (7.35) be continuous, bounded and satisfy the Lipschitz condition (5.18). Suppose $\varphi \in C_0^2(\mathbb{R}^n)$. Then the Cauchy problem for the time-fractional distributed order differential equation*

$$D_\mu u(t, x) = \mathcal{A}(x, D_x)u(t, x), \ t > 0, \ x \in \mathbb{R}^n,$$

$$u(0, x) = \varphi(x), \ x \in \mathbb{R}^n,$$

has a unique solution $u(t, x) \in C_0^2(\mathbb{R}^n)$ for each $t > 0$.

Proof. The result follows from the representation (7.37) in conjunction with estimate (7.17). □

Below we illustrate another method of derivation of time-fractional FPK equations associated with SDEs driven by a time-changed process. The discussion is based on the time-changed Itô formula (Theorem 6.1) and *does not use the Duality Theorem* (Theorem 6.2). For simplicity, we consider only the one-dimensional case with a time-changed Brownian motion with drift as the driving process. In this case the conjugate operator $\mathcal{A}^*$ corresponding to the forward FPK equation (7.6) is represented by

$$\mathcal{A}^*\varphi(y) = -\frac{\partial}{\partial y}\{b(y)\varphi(y)\} + \frac{1}{2}\frac{\partial^2}{\partial y^2}\{\sigma^2(y)\varphi(y)\}. \tag{7.40}$$

Theorem 7.6. *Let B_t be a standard $(\mathcal{F}_t)$-Brownian motion. Let $U_t = \sum_{k=1}^N c_k U_{k,t}$, where c_k are positive constants and $U_{k,t}$ are independent stable subordinators of respective indices $\beta_k \in (0,1)$. Let E_t be the inverse process to U_t. Suppose that X_t is a solution to the SDE*

$$dX_t = b(X_t)dE_t + \sigma(X_t)dB_{E_t}, \quad X_0 = x, \tag{7.41}$$

where $b(y)$ and $\sigma(y)$ satisfy the Lipschitz condition (5.18). Suppose also that X_{U_t} is independent of E_t. Then the transition probability density $p^X(t,x,y)$ satisfies in the weak sense the time-fractional partial differential equation

$$\sum_{k=1}^N c_k^{\beta_k} D_*^{\beta_k} p^X(t,x,y) = \mathcal{A}^* p^X(t,x,y), \quad (\mathcal{A}^* \text{ acts on variable } y) \tag{7.42}$$

with initial condition $p^X(0,x,y) = \delta_x(y)$, the Dirac delta function with mass on x. Here, $\mathcal{A}^*$ is the conjugate operator in (7.40).

Proof. For simplicity, the proof is given for $N = 2$. Recall that a process Z is said to be in synchronization with a time-change E (denoted as $Z \sim_{\text{synch}} E$) if Z is constant on every interval of the form $[E_{t-}, E_t]$ a.s. (see Sections 5.2 and 6.1).

Let $Y_t = X_{U_t}$. Since $E \sim_{\text{synch}} U$ and $B \circ E \sim_{\text{synch}} U$, it follows from Lemma 6.3 that $\int b(X)dE \sim_{\text{synch}} U$ and $\int \sigma(X)d(B \circ E) \sim_{\text{synch}} U$. Therefore, $X \sim_{\text{synch}} U$ due to SDE (7.41), yielding $Y_{E_t} = X_{U_{E(t)}} = X_t$. Hence, by the independence assumption between Y_t and E_t,

$$p^X(t,x,y) = \int_0^\infty p^Y(u,x,y)f_{E_t}(u)du \tag{7.43}$$

in the sense of distributions.

Since we are not assuming the Duality Theorem (Theorem 6.2), the fact that p^Y satisfies the classical FPK equation $\frac{\partial}{\partial t}p^Y(t,x,y) = \mathcal{A}^* p^Y(t,x,y)$ cannot be used here. Instead, we employ the time-changed Itô formula (Theorem 6.1) to obtain another representation of p^X in terms of p^Y as follows. Let $f \in C_c^\infty(\mathbb{R})$ (i.e. f is a

C^∞ function with compact support). Due to the fact that $X \sim_{\text{synch}} D$, it follows that $X_{D(s-)} = X_{D_s} = Y_s$ and the time-changed Itô formula (6.8) yields

$$f(X_t) - f(x) = \int_0^{E_t} f'(Y_s)b(Y_s)ds + \int_0^{E_t} f'(Y_s)\sigma(Y_s)dB_s \qquad (7.44)$$

$$+ \frac{1}{2}\int_0^{E_t} f''(Y_s)\sigma^2(Y_s)ds.$$

Because $f \in C_c^\infty(\mathbb{R})$, the process M defined by $M_u := \int_0^u f'(Y_s)\sigma(Y_s)dB_s$ is an $(\mathcal{F}_t)$-martingale. Taking expectations in (7.44) and conditioning on E_t which has density f_{E_t} given in (5.40),

$$\mathbb{E}[f(X_t)|X_0 = x] - f(x)$$

$$= \mathbb{E}\Big[M_{E_t} + \int_0^{E_t}\Big\{f'(Y_s)b(Y_s) + \frac{1}{2}f''(Y_s)\sigma^2(Y_s)\Big\}ds\Big|X_0 = x\Big]$$

$$= \int_0^\infty \mathbb{E}\Big[M_u + \int_0^u\Big\{f'(Y_s)b(Y_s) + \frac{1}{2}f''(Y_s)\sigma^2(Y_s)\Big\}ds\Big|E_t = u, Y_0 = x\Big]f_{E_t}(u)du$$

$$= \int_0^\infty \int_0^u \mathbb{E}\Big[f'(Y_s)b(Y_s) + \frac{1}{2}f''(Y_s)\sigma^2(Y_s)\Big|Y_0 = x\Big]ds\, f_{E_t}(u)du$$

by the assumption that $Y_t = X_{U_t}$ is independent of E_t. Using p^Y, the above can be rewritten as

$$\mathbb{E}[f(X_t)|X_0 = x] - f(x) \qquad (7.45)$$

$$= \int_0^\infty \int_0^u \int_{-\infty}^\infty \Big\{f'(y)b(y) + \frac{1}{2}f''(y)\sigma^2(y)\Big\}p^Y(s,x,y)dy\, ds\, f_{E_t}(u)du$$

$$= \int_{-\infty}^\infty f(y)\Big\{\int_0^\infty (J\mathcal{A}^*p^Y(u,x,y))f_{E_t}(u)du\Big\}dy,$$

where J is the integral operator. On the other hand, re-expressing the left-hand side of (7.45) in terms of p^X yields

$$\mathbb{E}[f(X_t)|X_0 = x] - f(x) = \int_{-\infty}^\infty f(y)p^X(t,x,y)dy - f(x). \qquad (7.46)$$

Since $f \in C_c^\infty(\mathbb{R})$ is arbitrary and $C_c^\infty(\mathbb{R})$ is dense in $L^2(\mathbb{R})$, comparison of (7.45) and (7.46) leads to another representation of p^X with respect to p^Y:

$$p^X(t,x,y) - \delta_x(y) = \int_0^\infty (J\mathcal{A}^*p^Y(u,x,y))f_{E_t}(u)du \qquad (7.47)$$

in the sense of distributions with $p^X(0,x,y) = \delta_x(y)$.

Now, we use the two representations (7.43) and (7.47) to derive equation (7.42) with the help of Laplace transforms. The Laplace transform of a function $v(t)$ of the form in (7.18), with f_{E_t} in (5.40), is computed as in (7.19). Using this fact and taking the Laplace transform of both sides in (7.43), we obtain

$$\widetilde{p^X}(s,x,y) = (C_1 s^{\beta_1-1} + C_2 s^{\beta_2-1})\widetilde{p^Y}(C_1 s^{\beta_1} + C_2 s^{\beta_2}, x, y), \quad s > 0,$$

where $C_k = c_k^{\beta_k} \, (k = 1, 2)$; whereas the Laplace transform of (7.47) is

$$\widetilde{p^X}(s, x, y) - \frac{1}{s}\delta_x(y) = (C_1 s^{\beta_1 - 1} + C_2 s^{\beta_2 - 1}) \, \widetilde{J A^* p^Y}(C_1 s^{\beta_1} + C_2 s^{\beta_2}, x, y)$$

$$= \frac{C_1 s^{\beta_1 - 1} + C_2 s^{\beta_2 - 1}}{C_1 s^{\beta_1} + C_2 s^{\beta_2}} \, \widetilde{A^* p^Y}(C_1 s^{\beta_1} + C_2 s^{\beta_2}, x, y), \ \ s > 0.$$

Combining these two identities,

$$C_1\big(s^{\beta_1}\widetilde{p^X}(s, x, y) - s^{\beta_1 - 1}\delta_x(y)\big) + C_2\big(s^{\beta_2}\widetilde{p^X}(s, x, y) - s^{\beta_2 - 1}\delta_x(y)\big)$$

$$= (C_1 s^{\beta_1} + C_2 s^{\beta_2})\Big(\widetilde{p^X}(s, x, y) - \frac{1}{s}\delta_x(y)\Big)$$

$$= (C_1 s^{\beta_1 - 1} + C_2 s^{\beta_2 - 1}) \, \widetilde{A^* p^Y}(C_1 s^{\beta_1} + C_2 s^{\beta_2}, x, y)$$

$$= \widetilde{A^* p^X}(s, x, y), \ \ s > 0,$$

which coincides with the identity obtained from applying the Laplace transform to both sides of (7.42). □

Remark 7.3. a) The proof given for Theorem 7.6 does not work if SDE (7.41) contains an additional term $\rho(X_t)dt$. In this case, the relationship $Y_{E_t} = X_t$ does not always follow from the definition $Y_t := X_{U_t}$. Consequently, identity (7.43) is not obtained. Example 5.4 in [Kobayashi (2011)] yields the following conjecture: if an additional term $\rho(X_t)dt$ is included in SDE (7.41), where $\rho(y)$ also satisfies the Lipschitz condition, then it is expected that the partial differential equation corresponding to (7.42) may involve a fractional integral term.

b) For a detailed analysis and concrete examples of SDEs driven by a time-changed Brownian motion of the form (7.41), consult Sections 4 and 5 of [Kobayashi (2011)].

7.2.2 Applications

Application 1 (Time-changed Lévy process). The operator $\mathcal{A}$ associated with a Lévy process L_t with characteristics (b, Σ, ν) is a pseudo-differential operator with the symbol $\Psi(\xi)$ given in (5.14). The corresponding Cauchy problem takes the form

$$\frac{\partial u(t, x)}{dt} = \mathcal{A}(D_x)u(t, x), \ u(0, x) = \varphi(x), \tag{7.48}$$

where $\mathcal{A}(D_x)$ is the pseudo-differential operator with the symbol $\Psi(\xi)$. Theorem 7.4 implies that if $E(\mu, t)$ is the inverse of $U(\mu, t) \in \mathcal{S}$ which is independent of L_t, then the Cauchy problem associated with the time-changed Lévy process L_{E_t} is the initial value problem for the *TDFPKΨ* equation

$$D_\mu u(t, x) = \mathcal{A}(D_x)u(t, x).$$

Application 2 (Time-changed α-stable Lévy process). Let $L_{\alpha, t}$ be a spherically symmetric α-stable Lévy process in $\mathbb{R}^n$, which is a pure jump process; see Section

5.3 for the definition. If $p^L(t,x) = E[\varphi(L_{\alpha,t})|L_{\alpha,0} = x]$, where $\varphi \in C_0^2(\mathbb{R}^n)$ (or $\varphi \in H^\alpha(\mathbb{R}^n)$), the Sobolev space of order α), then $p^L(t,x)$ satisfies in the strong sense the Cauchy problem

$$\frac{\partial p^L(t,x)}{\partial t} = -\kappa_\alpha(-\Delta)^{\alpha/2} p^L(t,x), \ t > 0, \ x \in \mathbb{R}^n, \tag{7.49}$$

$$p^L(0,x) = \varphi(x), \ x \in \mathbb{R}^n, \tag{7.50}$$

where κ_α is a constant depending on α and $-(-\Delta)^{\alpha/2}$ is a fractional power of the Laplace operator. The operator on the right hand side of (7.49) can be represented as a pseudo-differential operator with the symbol $\psi(\xi) = -|\xi|^\alpha$. As we observed in Section 3.7, it can also be represented as a hyper-singular integral, which is more convenient in random walk approximations of spherically symmetric α-stable Lévy processes discussed in Section 6.4.

Now suppose Y_t solves SDE

$$dY_t = g(Y_{t-})dL_{\alpha,t}, \ Y_0 = x, \tag{7.51}$$

where $g(x)$ is a Lipschitz-continuous function satisfying the linear growth condition. In this case, the forward Kolmogorov equation takes the form

$$\frac{\partial p^Y(t,y)}{\partial t} = -\kappa_\alpha(-\Delta)^{\alpha/2}\{[g(y)]^\alpha p^Y(t,y)\}, \ t > 0, \ y \in \mathbb{R}^n. \tag{7.52}$$

Application of Theorem 7.4 implies that $X_t = Y_{E(\mu,t)}$ satisfies the SDE

$$dX_t = g(X_{t-})dL_{\alpha,E(\mu,t)}, \ X_0 = x, \tag{7.53}$$

where $E(\mu,t)$ is the inverse of $U(\mu,t) \in \mathcal{S}$ which is independent of Y_t. Moreover, if $E(\mu,t)$ is independent of Y_t, then the corresponding forward Kolmogorov equation becomes a *TDFPKΨ* equation

$$D_\mu p^X(t,y) = -\kappa_\alpha(-\Delta)^{\alpha/2}\{[g(y)]^\alpha p^X(t,y)\}, \ t > 0, \ y \in \mathbb{R}^n, \tag{7.54}$$

where D_μ is the operator defined in (7.25). When the SDE in (7.53) is driven by a non-symmetric α-stable Lévy process, an analogue of (7.54) holds using instead of (7.52) its analogue appearing in [Schertzer et al. (2001)].

Application 3 (Fractional analogue of the Feynman–Kac formula). Suppose Y_t is a strong solution of SDE (7.30). Let $\bar{Y} \in \mathbb{R}^n$ be a fixed point, which we call a terminal point. Let q be a nonnegative continuous function. Consider the process

$$Y_t^q = \begin{cases} Y_t, & \text{if } 0 \leqslant t < \mathcal{T}_q, \\ \bar{Y}, & \text{if } t \geqslant \mathcal{T}_q, \end{cases}$$

where $\mathcal{T}_q$ is an $(\mathcal{F}_t)$-stopping time satisfying

$$\mathbb{P}(\mathcal{T}_q > t|\mathcal{F}_t) = \exp\left(-\int_0^t q(Y_s)ds\right).$$

The process Y_t^q is a Feller process with associated semigroup (see [Applebaum (2009)])

$$(T_t^q \varphi)(x) = \mathbb{E}\left[\exp\left(-\int_0^t q(Y_s)ds\right)\varphi(Y_t)\Big|Y_0 = x\right] \qquad (7.55)$$

and infinitesimal generator $\mathcal{A}_q(x, D_x) = -q(x) + \mathcal{A}(x, D_x)$, where $\mathcal{A}(x, D_x)$ is the pseudo-differential operator defined in (7.34). Let E_t^β be the inverse to a β-stable subordinator independent of Y_t. Then it follows from Theorems 7.1 and 7.3 with $N = 1$ that $X_t := Y_{E_t^\beta}$ solves SDE (7.31) and the function

$$u(t, x) = \mathbb{E}\left[\exp\left(-\int_0^t q(X_s)dE_s^\beta\right)\varphi(X_t)\Big|X_0 = x\right] \qquad (7.56)$$

solves the Cauchy problem for the fractional order equation

$$D_*^\beta u(t, x) = [-q(x) + \mathcal{A}(x, D_x)]u(t, x), \ t > 0, \ x \in \mathbb{R}^n,$$
$$u(0, x) = \varphi(x), \ x \in \mathbb{R}^n.$$

Formula (7.56) represents a fractional analogue of the Feynman–Kac formula.

7.3 FPK equations associated with SDEs driven by fractional Brownian motion

This section provides FPK equations associated with SDEs driven by i) fractional Brownian motion and ii) a time-changed (non-fractional) Brownian motion. For ii), we present an operator approach, which will be used in Section 7.4 to derive time-fractional FPK equations associated with stochastic processes which are time changes of solutions of SDEs driven by fractional Brownian motion.

Recall that one-dimensional fractional Brownian motion (fBM) B_t^H is a zero-mean Gaussian process with continuous paths and covariance function

$$R_H(s, t) = \mathbb{E}[B_s^H B_t^H] = \frac{1}{2}(s^{2H} + t^{2H} - |s - t|^{2H}), \qquad (7.57)$$

where the Hurst parameter H takes values in $(0, 1)$. If $H = 1/2$, then B_t^H becomes a standard Brownian motion. Fractional Brownian motion B_t^H, like standard Brownian motion, has nowhere differentiable sample paths and stationary increments, but it does not have independent increments unless $H = 1/2$. More properties of fBM are provided in Section 5.5.

Fractional Brownian motion is not a semimartingale unless $H = 1/2$ ([Biagini et al. (2008), Nualart (2006)]), so the usual Itô's stochastic calculus is not valid. Nevertheless, there are several approaches [Bender (2003), Biagini et al. (2008), Decreusefond and Uštünel (1998), Nualart (2006)] to a stochastic calculus in order to interpret in a meaningful way an SDE driven by an m-dimensional fBM B_t^H of the form

$$Y_t = Y_0 + \int_0^t b(Y_s)ds + \int_0^t \sigma(Y_s)dB_s^H, \qquad (7.58)$$

where mappings $b : \mathbb{R}^n \to \mathbb{R}^n$ and $\sigma : \mathbb{R}^n \to \mathbb{R}^{n \times m}$ are Lipschitz continuous and bounded; Y_0 is a random variable independent of B_t^H. We do not discuss these approaches here, referring the interested reader to [Biagini et al. (2008), Coutin and Decreusefond (1997), Nualart (2006)]. Instead, we focus our attention on the FPK equation associated with SDE (7.58) driven by fBM whose generic form is given by

$$\frac{\partial u(t,x)}{\partial t} = B(x, D_x)u(t,x) + Ht^{2H-1}A(x, D_x)u(t,x), \tag{7.59}$$

where $B(x, D_x) = \sum_{j=1}^n b_j(x)\frac{\partial}{\partial x_j}$, a first order differential operator, and $A(x, D_x)$ is a second order elliptic differential operator

$$A(x, D_x) = \sum_{j,k=1}^n a_{jk}(x)\frac{\partial^2}{\partial x_j \partial x_k}. \tag{7.60}$$

Functions $a_{jk}(x)$, $j, k = 1, \ldots, n$ are entries of the matrix $\mathcal{A}(x) = \sigma(x) \times \sigma^T(x)$, where $\sigma^T(x)$ is the transpose of matrix $\sigma(x)$. By definition $\mathcal{A}(x)$ is positive definite: for any $x \in \mathbb{R}^n$ and $\xi \in \mathbb{R}^n$ one has $\sum_{j,k=1}^n a_{jk}(x)\xi_j\xi_k \geq C|\xi|^2$, where C is a positive constant. The operator $A(x, D_x)$ can also be given in the divergent form

$$A(x, D_x) = \sum_{j,k=1}^n \frac{\partial}{\partial x_j}\left(a_{jk}(x)\frac{\partial}{\partial x_k}\right). \tag{7.61}$$

The right hand side of (7.59) depends on the time variable t, which, in fact, reflects the presence of correlation. Additionally, $u(t,x)$ in equation (7.59) satisfies the initial condition

$$u(0, x) = \varphi(x), \quad x \in \mathbb{R}^n, \tag{7.62}$$

where $\varphi(x)$ belongs to some function space, or is a generalized function (in which case the solution to the FPK equation is understood in the weak sense).

In the one-dimensional case with $H \in (1/4, 1)$, Example 28 in [Baudoin and Coutin (2007)] establishes that if Y_t solves SDE (7.58) with $b = 0$, where the stochastic integral is understood in the sense of Stratonovich, then $u(t,x) = \mathbb{E}[\varphi(Y_t)|Y_0 = x]$ solves the FPK equation (7.59) with initial condition (7.62), where the operator $A(x, D_x)$ is expressed in the divergence form (7.61). Paper [Gazanfer (2006)] derives an FPK equation with $A(x, D_x)$ in the form (7.60). However, the derivation of the FPK equation in this paper is based on the Itô formula obtained in [Bender (2003)] for fBM (that is for $b = 0$ and $\sigma = I$, the identity matrix). Therefore, their derivation might require a modification. In the general setting of (7.58) and (7.59), it is not known to us whether $u(t,x) = \mathbb{E}[\varphi(Y_t)|Y_0 = x]$ solves (7.59) with initial condition (7.62) when Y_t solves (7.58).

In the sequel we use the following notation:

$$L_\gamma(t, x, D_x) = B(x, D_x) + \frac{\gamma+1}{2}t^\gamma A(x, D_x), \tag{7.63}$$

where $\gamma = 2H - 1$. The introduction of γ is made so that the operators G_γ arising in Section 7.4 will have the semigroup property. If $\gamma = 0$, equivalently $H = 1/2$, then the operator $L_0(t, x, D_x)$ has a form with coefficients not depending on t:

$$L_0(t, x, D_x) \equiv L(x, D_x) = B(x, D_x) + \frac{1}{2}A(x, D_x), \tag{7.64}$$

and equation (7.59) takes the form

$$\frac{\partial u(t,x)}{\partial t} = L(x, D_x)u(t, x), \tag{7.65}$$

which coincides with the FPK equation (7.1) for the classical diffusion.

Let E^β be a time-change given by the inverse of a stable subordinator U^β with index $\beta \in (0, 1)$, independent of B_t^H. Suppose $H = 1/2$ and the stochastic integral in SDE (7.58) is understood in the sense of Itô. If the driving process in (7.58) is replaced by a composition of the driving process with E^β, then as we observed in Theorems 7.3 and 7.6 in Section 7.2, the left hand side of equation (7.65) becomes a fractional derivative of order β and the right hand side remains unchanged:

$$D_*^\beta v(t,x) = L(x, D_x)v(t, x). \tag{7.66}$$

As we will see in Section 7.4, however, *this is not the case for fractional FPK equations associated with time-changed fBM with $H \neq 1/2$.* This is essentially a consequence of the correlation of the increments of fBM. Note also that since fBM with $H \neq 1/2$ is not a semimartingale, the method used in Section 7.2 is not applicable in this case. The time-changed fBM appears as a continuous time random walk (CTRW) scaling limit for certain correlated random variables; see [Meerschaert et al. (2009a)]. Authors of that paper write, "An interesting open question is to establish the governing equation for the CTRW scaling limit." A particular case of Theorem 7.8 in Section 7.4 answers that question. (See also Remark 7.7 (c).)

Remark 7.4. It is unknown how to interpret an SDE driven by a time-changed fBM with $H \neq 1/2$ of the form

$$X_t = x_0 + \int_0^t b(X_s)dE_s + \int_0^t \sigma(X_s)dB_{E_s}^H \tag{7.67}$$

in such a way that a solution X_t is represented as $X_t = Y_{E_t}$ with Y_t solving SDE (7.58). Such an interpretation together with Theorems 7.8 and 7.9 in Section 7.4 would allow us to generalize the relationship (i)–(ii) in Section 7.1 to a class of FPK equations that is different from the class of *TFFPKΨ/TDFPKΨ* equations discussed in Section 7.2.

7.3.1 *An operator approach to derivation of fractional FPK equations*

There are several approaches for deriving equation (7.66) including via semigroup theory [Bauemer and Meerschaert (2001), Meerschaert et al. (2002a), Meerschaert and Scheffler (2004), Hahn et al. (2012)], master equations [Meerschaert et al. (2002b), Scalas et al. (2000)], and CTRWs [Gorenflo and Mainardi (1998), Gorenflo and Mainardi (2008), Meerschaert and Scheffler (2008), Umarov and Steinberg (2006)]. In the remainder of this section, we present a different technique which can be extended for equations with a time-dependent right hand side as well, including equations of the form (7.59). This technique is close to the method used in [Kolokoltsov (2009)].

Notice that solutions to equations (7.66) and (7.65) are connected by a certain relationship. Namely, a solution $v(t, x)$ to equation (7.66) satisfying the initial condition (7.62) can be represented through the solution $u(t, x)$ to equation (7.65) satisfying the same initial condition (7.62) via the formula

$$v(t, x) = \int_0^\infty f_{E_t^\beta}(\tau) u(\tau, x) d\tau, \tag{7.68}$$

where $f_{E_t^\beta}$ is the density function of E_t^β for each fixed $t > 0$. Let $f_{U_1^\beta}$ be the density function of the corresponding β-stable subordinator U_t^β at $t = 1$. Then by (5.32),

$$f_{E_t^\beta}(\tau) = -\frac{\partial}{\partial \tau}[Jf_{U_1^\beta}]\left(\frac{t}{\tau^{1/\beta}}\right) = \frac{t}{\beta\tau^{1+1/\beta}} f_{U_1^\beta}\left(\frac{t}{\tau^{1/\beta}}\right), \tag{7.69}$$

where $Jf(t) = \int_0^t f(u) du$ is the integration operator. Since $f_{U_1^\beta}(u) \in C^\infty(0, \infty)$, it follows from representation (7.69) that $f_{E_t^\beta}(\tau) \in C^\infty((0, \infty) \times (0, \infty))$. Further properties of $f_{E_t^\beta}(\tau)$ are presented in the following lemma.

Lemma 7.1. *Let $f_{E_t^\beta}(\tau)$ be the function given in (7.69). Then*

(a) $\lim_{t \to +0} f_{E_t^\beta}(\tau) = \delta_0(\tau)$ *in the sense of the topology of the space of tempered distributions $\mathcal{D}'(\mathbb{R})$;*

(b) $\lim_{\tau \to +0} f_{E_t^\beta}(\tau) = \dfrac{t^{-\beta}}{\Gamma(1 - \beta)}$, $t > 0$;

(c) $\lim_{\tau \to \infty} f_{E_t^\beta}(\tau) = 0$, $t > 0$;

(d) $L_{t \to s}[f_{E_t^\beta}(\tau)](s) = s^{\beta-1} e^{-\tau s^\beta}$, $s > 0$, $\tau \geqslant 0$,

where $L_{t \to s}$ denotes the Laplace transform with respect to the variable t.

Proof. (a) Let $\psi(\tau)$ be an infinitely differentiable function rapidly decreasing at infinity. We have to show that $\lim_{t \to +0}\langle f_{E_t^\beta}, \psi \rangle = \psi(0)$. Here $\langle f_{E_t^\beta}, \psi \rangle$ denotes the value of $f_{E_t^\beta} \in \mathcal{D}'(\mathbb{R})$ on ψ. We have

$$\lim_{t \to +0}\langle f_{E_t^\beta}(\tau), \psi(\tau) \rangle = \lim_{t \to +0} \int_0^\infty f_{E_t^\beta}(\tau)\psi(\tau) d\tau = \lim_{t \to +0} \int_0^\infty f_{U_1^\beta}(u)\psi\left(\left(\frac{t}{u}\right)^\beta\right) du$$

$$= \psi(0) \int_0^\infty f_{U_1^\beta}(u) du = \psi(0).$$

Parts (b) and (c) follow from asymptotic relations (5.28) and (5.27), respectively. Part (d) is straightforward; we only need to compute the Laplace transform of $f_{E_t^\beta}(\tau)$ using the representation $f_{E_t^\beta}(\tau) = -\frac{\partial}{\partial \tau}[Jf_{U_1^\beta}](\frac{t}{\tau^{1/\beta}})$. Recall the following property of the Laplace transform

$$L[Jf](s) = \frac{1}{s}L[f](s), \tag{7.70}$$

which implies that the relation

$$L[Jf(t/a)](s) = \frac{1}{s}L[f](as) \tag{7.71}$$

is valid for arbitrary $a > 0$. Indeed,

$$L[Jf(t/a)](s) = \int_0^\infty e^{-st}\left[\int_0^{t/a} f(u)du\right]dt = a\int_0^\infty e^{-asv}\left[\int_0^v f(u)du\right]dv$$

$$= aL[Jf(t)](as) = \frac{a}{as}L[f](as) = \frac{1}{s}L[f](as),$$

yielding (7.71). Now using (7.71) with $a = \tau^{1/\beta}$,

$$L_{t\to s}[f_{E_t^\beta}(\tau)](s) = L_{t\to s}\left[-\frac{\partial}{\partial\tau}Jf_{U_1^\beta}(\frac{t}{\tau^{1/\beta}})\right](s)$$

$$= -\frac{\partial}{\partial\tau}L_{t\to s}\left[Jf_{U_1^\beta}(\frac{t}{\tau^{1/\beta}})\right](s)$$

$$= -\frac{1}{s}\frac{\partial}{\partial\tau}L_{t\to s}\left[f_{U_1^\beta}(t)\right](\tau^{1/\beta}s) = -\frac{1}{s}\frac{\partial}{\partial\tau}\left[e^{-\tau s^\beta}\right]$$

$$= s^{\beta-1}e^{-\tau s^\beta},$$

proving part (d). $\qquad\square$

We note that due to part (b) of Lemma 7.1, $f_{E_t^\beta} \in C^\infty(0,\infty)$ for each fixed $\tau \geq 0$. Hence, the fractional derivative $D_{*,t}^\beta f_{E_t^\beta}(\tau)$ in the variable t is meaningful and is a generalized function of the variable τ.

Lemma 7.2. *The function $f_{E_t^\beta}(\tau)$ defined in (7.69) for each $t > 0$ satisfies the equation*

$$D_{*,t}^\beta f_{E_t^\beta}(\tau) = -\frac{\partial}{\partial\tau}f_{E_t^\beta}(\tau) - \frac{t^{-\beta}}{\Gamma(1-\beta)}\delta_0(\tau) \qquad (7.72)$$

in the sense of tempered distributions.

Proof. The Laplace transform (in variable t) of the left hand side, using the definition (7.69) of $f_{E_t^\beta}(\tau)$, equals

$$L_{t\to s}[D_{*,t}^\beta f_{E_t^\beta}(\tau)](s) = s^\beta L_{t\to s}[f_{E_t^\beta}(\tau)](s) - s^{\beta-1}\lim_{t\to+0}f_{E_t^\beta}(\tau)$$

$$= s^{2\beta-1}e^{-\tau s^\beta} - s^{\beta-1}\delta_0(\tau), \quad s > 0.$$

On the other hand, the Laplace transform of the right hand side equals

$$L_{t\to s}\left[-\frac{\partial}{\partial\tau}f_{E_t^\beta}(\tau) - \frac{t^{-\beta}}{\Gamma(1-\beta)}\delta_0(\tau)\right](s) = \frac{\partial^2}{\partial\tau^2}\left(\frac{1}{s}e^{-\tau s^\beta}\right) - s^{\beta-1}\delta_0(\tau)$$

$$= s^{2\beta-1}e^{-\tau s^\beta} - s^{\beta-1}\delta_0(\tau), \quad s > 0,$$

completing the proof. $\qquad\square$

Now it is easy to derive the fractional order FPK equation (7.66), a solution of which is given by $v(t,x)$ in (7.68). We have

$$
\begin{aligned}
D_{*,t}^{\beta} v(t,x) &= \int_0^{\infty} D_{*,t}^{\beta} f_{E_t^{\beta}}(\tau) u(\tau,x) d\tau \\
&= -\int_0^{\infty} \left[\frac{\partial}{\partial \tau} f_{E_t^{\beta}}(\tau) + \frac{t^{-\beta}}{\Gamma(1-\beta)} \delta_0(\tau) \right] u(\tau,x) d\tau \\
&= -\lim_{\tau \to \infty} \left[f_{E_t^{\beta}}(\tau) u(\tau,x) \right] + \lim_{\tau \to 0} \left[f_{E_t^{\beta}}(\tau) u(\tau,x) \right] \\
&\quad + \int_0^{\infty} f_{E_t^{\beta}}(\tau) \frac{\partial}{\partial \tau} u(\tau,x) d\tau - \frac{t^{-\beta}}{\Gamma(1-\beta)} u(0,x).
\end{aligned}
$$

Due to Lemma 7.1, part (c) implies the first term vanishes since $u(\tau,x)$ is bounded, while part (b) implies the second and last terms cancel. Taking into account (7.65),

$$
D_{*,t}^{\beta} v(t,x) = \int_0^{\infty} f_{E_t^{\beta}}(\tau) L(x, D_x) u(\tau,x) d\tau = L(x, D_x) v(t,x). \tag{7.73}
$$

Moreover, by property (a) of Lemma 7.1,

$$
\lim_{t \to +0} v(t,x) = \langle \delta_0(\tau), u(\tau,x) \rangle = u(0,x).
$$

The above technique extends to the more general case when the time-change is the inverse of an arbitrary mixture of independent stable subordinators. As in Section 7.2, let $U(\mu,t) \in \mathcal{S}$ denote a strictly increasing subordinator satisfying

$$
\mathbb{E}[e^{-sU(\mu,t)}] = e^{-t\Phi_\mu(s)} \quad \text{with} \quad \Phi_\mu(s) = \int_0^1 s^{\beta} d\mu(\beta)
$$

for some finite positive measure μ on $(0,1)$.

Theorem 7.7. *Let $u(t,x)$ be a solution of the Cauchy problem*

$$
\frac{\partial u(t,x)}{\partial t} = L(x, D_x) u(t,x), \ t > 0, \ x \in \mathbb{R}^n, \tag{7.74}
$$

$$
u(0,x) = \varphi(x), \ x \in \mathbb{R}^n. \tag{7.75}
$$

Let $E(\mu,t)$ be the inverse of a subordinator $U(\mu,t) \in \mathcal{S}$. Then the function $v(t,x) = \int_0^{\infty} f_{E(\mu,t)}(\tau) u(\tau,x) d\tau$, where $f_{E(\mu,t)}$ is the density function of $E(\mu,t)$, satisfies the initial value problem for the distributed order fractional differential equation

$$
D_\mu v(t,x) \equiv \int_0^1 D_{*,t}^{\beta} v(t,x) d\mu(\beta) = L(x, D_x) v(t,x), \ t > 0, \ x \in \mathbb{R}^n, \tag{7.76}
$$

$$
v(0,x) = \varphi(x), \ x \in \mathbb{R}^n. \tag{7.77}
$$

The proof of this theorem requires two lemmas which generalize Lemmas 7.1 and 7.2. Define the function

$$
\mathcal{K}_\mu(t) = \int_0^1 \frac{t^{-\beta}}{\Gamma(1-\beta)} d\mu(\beta), \ t > 0. \tag{7.78}
$$

Lemma 7.3. *Let $f_{E(\mu,t)}(\tau)$ be the function defined in Theorem 7.7. Then*

(a) $\lim\limits_{t \to +0} f_{E(\mu,t)}(\tau) = \delta_0(\tau)$, $\tau \geqslant 0$;

(b) $\lim\limits_{\tau \to +0} f_{E(\mu,t)}(\tau) = \mathcal{K}_\mu(t)$, $t > 0$;

(c) $\lim\limits_{\tau \to \infty} f_{E(\mu,t)}(\tau) = 0$, $t > 0$;

(d) $L_{t \to s}[f_{E(\mu,t)}(\tau)](s) = \dfrac{\Phi_\mu(s)}{s} e^{-\tau \Phi_\mu(s)}$, $s > 0$, $\tau \geqslant 0$.

Proof. First notice that $f_{E(\mu,t)}(\tau) = -\frac{\partial}{\partial \tau}[J f_{U(\mu,\tau)}](t)$, where J is the usual integration operator. The proofs of parts (a)–(c) are similar to the proofs of parts (a)–(c) of Lemma 7.1. Further, using the definition of $U(\mu,t)$,

$$L_{t \to s}[f_{E(\mu,t)}(\tau)](s) = -\frac{1}{s}\frac{\partial}{\partial \tau} L_{t \to s}[f_{U(\mu,\tau)}(t)](s) = \frac{\Phi_\mu(s)}{s} e^{-\tau \Phi_\mu(s)}, \ s > 0,$$

which completes the proof. $\qquad\square$

Lemma 7.4. *The function $f_{E(\mu,t)}(\tau)$ defined in Theorem 7.7 satisfies for each $t > 0$ the equation*

$$D_{\mu,t} f_{E(\mu,t)}(\tau) = -\frac{\partial}{\partial \tau} f_{E(\mu,t)}(\tau) - \delta_0(\tau)\mathcal{K}_\mu(t) \tag{7.79}$$

in the sense of tempered distributions, where $D_\mu = D_{\mu,t}$ is the distributed fractional order derivative defined in (7.76).

Proof. Integrating both sides of the equation

$$L_{t \to s}[D^\beta_{*,t} f_{E(\mu,t)}(\tau)] = s^\beta L_{t \to s}[f_{E(\mu,t)}(\tau)](s) - s^{\beta-1}\delta_0(\tau)$$

with respect to μ and taking into account part (d) of Lemma 7.3 yields

$$L_{t \to s}[D_{\mu,t} f_{E(\mu,t)}(\tau)] = \frac{\Phi_\mu^2(s)}{s} e^{-\tau \Phi_\mu(s)} - \frac{\Phi_\mu(s)}{s}\delta_0(\tau).$$

It is easy to verify that the latter coincides with the Laplace transform of the right hand side of (7.79). $\qquad\square$

Proof of Theorem 7.7. By Lemma 7.4,

$$
\begin{aligned}
D_{\mu,t} v(t,x) &= \int_0^\infty D_{\mu,t} f_{E(\mu,t)}(\tau) u(\tau,x) d\tau \\
&= -\lim_{\tau \to \infty}[f_{E(\mu,t)}(\tau) u(\tau,x)] + \lim_{\tau \to 0}[f_{E(\mu,t)}(\tau) u(\tau,x)] \\
&\quad + \int_0^\infty f_{E(\mu,t)}(\tau)\frac{\partial}{\partial \tau} u(\tau,x) d\tau - \mathcal{K}_\mu(t) u(0,x) \\
&= \int_0^\infty f_{E(\mu,t)}(\tau)\frac{\partial}{\partial \tau} u(\tau,x) d\tau,
\end{aligned}
$$

since all the limit expressions vanish due to parts (b) and (c) of Lemma 7.3. Now taking into account equation (7.74),

$$D_{\mu,t} v(t,x) = \int_0^\infty f_{E(\mu,t)}(\tau) L(x, D_x) u(\tau,x) d\tau = L(x, D_x) v(t,x). \tag{7.80}$$

The initial condition (7.77) is also verified by using property (a) of Lemma 7.3:

$$\lim_{t \to +0} v(t, x) = \langle \delta_0(\tau), u(\tau, x) \rangle = u(0, x) = \varphi(x).$$

This completes the proof.   □

Remark 7.5. The equivalent version of formula (7.73) in terms of Riemann-Liouville fractional derivatives was proven in [Bauemer and Meerschaert (2001)] in a more general setting. Theorem 7.7, Lemma 7.3 (d), and Lemma 7.4 are special cases of theorems and equations proven in [Meerschaert and Scheffler (2008)]. However, the above proofs are simpler and significantly different from the treatment in [Meerschaert and Scheffler (2008)].

7.4   Fractional FPK equations associated with stochastic processes which are time changes of solutions of SDEs driven by fractional Brownian motion

This section employs the operator approach established in Section 7.3 to derive fractional FPK equations for time-changed stochastic processes of the form Y_{E_t}, where Y_t solves an SDE driven by fBM. In particular, we derive fractional FPK equations associated with a time-changed fBM $B_{E_t}^H$ with $H \neq 1/2$, where the time-change E_t is independent of the fBM B_t^H. Recall that a generic form of the FPK equation associated with an SDE driven by an fBM (without time-change) is given by

$$\frac{\partial u(t, x)}{\partial t} = L_\gamma(t, x, D_x)u(t, x), \tag{7.81}$$

where $L_\gamma(t, x, D_x)$ is defined in (7.63) and the Hurst parameter H is connected with γ via $2H - 1 = \gamma$.

7.4.1   *Theory*

For simplicity, we first consider a time-change E_t^β inverse to a single stable subordinator U_t^β. Hence, the density function $f_{E_t^\beta}(\tau)$ of E_t^β possesses all the properties mentioned in Lemmas 7.1 and 7.2.

Theorem 7.8. *Let $u(t, x)$ be a solution to the initial value problem*

$$\frac{\partial u(t, x)}{\partial t} = B(x, D_x)u(t, x) + \frac{\gamma + 1}{2}t^\gamma A(x, D_x)u(t, x), \ t > 0, \ x \in \mathbb{R}^n, \tag{7.82}$$

$$u(0, x) = \varphi(x), \ x \in \mathbb{R}^n. \tag{7.83}$$

Let E_t^β be the inverse of a β-stable subordinator U_t^β with index $\beta \in (0, 1)$. Then the function $v(t, x) = \int_0^\infty f_{E_t^\beta}(\tau)u(\tau, x)d\tau$ satisfies the following initial value problem for a fractional order differential equation

$$D_*^\beta v(t, x) = B(x, D_x)v(t, x) + \frac{\gamma + 1}{2}G_{\gamma, t}A(x, D_x)v(t, x), \ t > 0, \ x \in \mathbb{R}^n, \tag{7.84}$$

$$v(0, x) = \varphi(x), \ x \in \mathbb{R}^n, \tag{7.85}$$

where the operator $G_{\gamma,t}$ acts on the variable t and is defined by

$$G_{\gamma,t}v(t,x) = \beta\Gamma(\gamma+1)J_t^{1-\beta}L_{s\to t}^{-1}\left[\frac{1}{2\pi i}\int_{C-i\infty}^{C+i\infty}\frac{\tilde{v}(z,x)}{(s^\beta - z^\beta)^{\gamma+1}}dz\right](t), \qquad (7.86)$$

with $0 < C < s$, and $z^\beta = e^{\beta Ln(z)}$, $Ln(z)$ being the principal value of the complex $\ln(z)$ with cut along the negative real axis. Here, the operators $J_t^{1-\beta}$ and $L_{s\to t}^{-1}$ represent the fractional integral of order $1 - \beta$ and the inverse Laplace transform, respectively.

Proof. Using the properties of $f_{E_t^\beta}(\tau)$, we obtain as in the proof of (7.73),

$$D_{*,t}^\beta v(t,x) = B(x,D_x)v(t,x) + \frac{\gamma+1}{2}A(x,D_x)G_{\gamma,t}v(t,x),$$

where

$$G_{\gamma,t}v(t,x) = \int_0^\infty f_{E_t^\beta}(\tau)\tau^\gamma u(\tau,x)d\tau. \qquad (7.87)$$

It follows from the definition (7.68) of $v(t,x)$ that if $\gamma = 0$, then $G_{0,t}$ is the identity operator. To show representation (7.86) in the case $\gamma \neq 0$, we find the Laplace transform of $G_{\gamma,t}v(t,x)$. By the property (d) of Lemma 7.1,

$$L[G_{\gamma,t}v(t,x)](s) = s^{\beta-1}\int_0^\infty e^{-\tau s^\beta}\tau^\gamma u(\tau,x)d\tau = s^{\beta-1}L[\tau^\gamma u(\tau,x)](s^\beta).$$

Obviously, if $\gamma = 0$, then $L[G_{0,t}v(t,x)](s) = s^{\beta-1}\tilde{u}(s^\beta,x)$, which implies $\tilde{v}(s,x) = s^{\beta-1}\tilde{u}(s^\beta,x)$. If $\gamma \neq 0$, then

$$L[t^\gamma u(t,x)](s) = L[t^\gamma](s) * \tilde{u}(s,x) = \frac{1}{2\pi i}\int_{c-i\infty}^{c+i\infty}\frac{\Gamma(\gamma+1)}{(s-z)^{\gamma+1}}\tilde{u}(z,x)dz, \qquad (7.88)$$

where $*$ stands for the convolution of Laplace images of two functions and $0 < c < s$. Now using the substitution $z = e^{\beta Ln(\zeta)}$, with $Ln(\zeta)$ the principal part of the complex function $\ln(\zeta)$, the right hand side of (7.88) reduces to

$$L[t^\gamma u(t,x)](s) = \frac{\beta}{2\pi i}\int_{C-i\infty}^{C+i\infty}\frac{\Gamma(\gamma+1)}{(s-\zeta^\beta)^{\gamma+1}}\zeta^{\beta-1}\tilde{u}(\zeta^\beta,x)d\zeta \qquad (7.89)$$

$$= \frac{\beta}{2\pi i}\int_{C-i\infty}^{C+i\infty}\frac{\Gamma(\gamma+1)}{(s-\zeta^\beta)^{\gamma+1}}\tilde{v}(\zeta,x)d\zeta.$$

The last equality uses the relation $\tilde{v}(\zeta,x) = \zeta^{\beta-1}\tilde{u}(\zeta^\beta,x)$. Further, replacing s by s^β and taking the inverse Laplace transform in (7.89) yields the desired representation (7.86) for the operator $G_{\gamma,t}$ since $L[J^{1-\beta}f](s) = s^{\beta-1}\tilde{f}(s)$. It follows from part (a) of Lemma 7.1 that $v(0,x) = u(0,x)$, which completes the proof. $\square$

In the more general case when the time-change process $E(\mu,t)$ is the inverse to $U(\mu,t) \in \mathcal{S}$, a mixture of independent stable subordinators with the mixing measure μ, a representation for the FPK equation is given in the following theorem. The proof is based on the properties of the density $f_{E(\mu,t)}(\tau)$ of $E(\mu,t)$ obtained in Lemmas 7.3 and 7.4.

Theorem 7.9. *Let $u(t,x)$ be a solution to the initial value problem (7.82)–(7.83). Let $E(\mu,t)$ be the inverse of a subordinator $U(\mu,t) \in S$. Then the function $v(t,x) = \int_0^\infty f_{E(\mu,t)}(\tau)u(\tau,x)d\tau$ satisfies the following initial value problem for a fractional order differential equation*

$$D_\mu v(t,x) = B(x,D_x)v(t,x) + \frac{\gamma+1}{2}G^\mu_{\gamma,t}A(x,D_x)v(t,x), \ t > 0, \ x \in \mathbb{R}^n, \quad (7.90)$$

$$v(0,x) = \varphi(x), \ x \in \mathbb{R}^n. \quad (7.91)$$

The operator $G^\mu_{\gamma,t}$ acts on the variable t and is defined by

$$G^\mu_{\gamma,t}v(t,x) = \mathcal{K}_\mu(t) * L^{-1}_{s\to t}\left[\frac{\Gamma(\gamma+1)}{2\pi i}\int_{C-i\infty}^{C+i\infty}\frac{m_\mu(z)\tilde{v}(z,x)}{(\Phi_\mu(s)-\Phi_\mu(z))^{\gamma+1}}dz\right](t), \quad (7.92)$$

where $*$ denotes the usual convolution of two functions, $0 < C < s$, $\Phi_\mu(z) = \int_0^1 e^{\beta Ln(z)}d\mu(\beta)$, $m_\mu(z) = \frac{1}{\Phi_\mu(z)}\int_0^1 \beta z^\beta d\mu(\beta)$, and $\mathcal{K}_\mu(t)$ is defined in (7.78).

Proof. The proof is similar to the proof of Theorem 7.8. We only sketch how to obtain representation (7.92) for the operator $G^\mu_{\gamma,t}v(t,x) = \int_0^\infty f_{E(\mu,t)}(\tau)\tau^\gamma u(\tau,x)d\tau$. The Laplace transform of $G^\mu_{\gamma,t}v(t,x)$, due to part (d) of Lemma 7.3, is

$$L_{t\to s}[G^\mu_{\gamma,t}v(t,x)](s) = \frac{\Phi_\mu(s)}{s}L[t^\gamma u(t,x)](\Phi_\mu(s)), \ s > 0.$$

Since $L[\mathcal{K}_\mu(t)](s) = \frac{\Phi_\mu(s)}{s}$, $s > 0$, we have

$$G^\mu_{\gamma,t}v(t,x) = \mathcal{K}_\mu(t) * L^{-1}_{s\to t}[L[t^\gamma u(t,x)](\Phi_\mu(s))](t).$$

Further, replacing s by $\Phi_\mu(s)$ in (7.88), followed by the substitution $z = \Phi_\mu(\zeta) = \int_0^1 e^{\beta Ln(\zeta)}d\mu(\beta)$ in the integral on the right side of (7.88), yields the form (7.92). $\square$

The following theorem represents the general case when the time-change E_t is not necessarily the inverse of a stable subordinator or their mixtures.

Theorem 7.10. *Let $\gamma \in (-1,1)$. Let E_t be a time-change whose density function $K(t,\tau) = f_{E_t}(\tau)$ satisfies the following hypotheses:*

i) $\lim_{\tau\to+0}[K(t,\tau)\tau^{-\gamma}] < \infty$ *for all $t > 0$;*

ii) $\lim_{\tau\to\infty}[K(t,\tau)\tau^{-\gamma}u(\tau,x)] = 0$ *for all $t > 0$ and $x \in \mathbb{R}^n$,*

where $u(t,x)$ is a solution to the initial value problem (7.82)–(7.83). Let H_t be an operator acting in the variable t such that

$$H_t K(t,\tau) = -\frac{\partial}{\partial\tau}\left[K(t,\tau)\left(\frac{t}{\tau}\right)^\gamma\right] - \delta_0(\tau)\lim_{\tau\to+0}\left[\left(\frac{t}{\tau}\right)^\gamma K(t,\tau)\right]. \quad (7.93)$$

Then the function $v(t, x) = \int_0^\infty K(t, \tau)u(\tau, x)d\tau$ *satisfies the initial value problem*

$$H_t v(t, x) = t^\gamma \bar{G}_{-\gamma,t} B(x, D_x) v(t, x)$$

$$+ \frac{\gamma + 1}{2} t^\gamma A(x, D_x) v(t, x), \quad t > 0, \ x \in \mathbb{R}^n, \tag{7.94}$$

$$v(0, x) = u(0, x), \ x \in \mathbb{R}^n, \tag{7.95}$$

where $\bar{G}_{-\gamma,t} v(t, x) = \int_0^\infty K(t, \tau)\tau^{-\gamma}u(\tau, x)d\tau.$

Remark 7.6. Obviously, if $\gamma \neq 0$, then H_t cannot be a fractional derivative in the sense of Caputo–Djrbashian (or Riemann–Liouville). A representation of H_t in cases when E_t is the inverse to a stable subordinator is given below in Corollary 7.11.

Proof. We have

$$H_t v(t, x) = \int_0^\infty H_t K(t, \tau)u(\tau, x)d\tau$$

$$= -\int_0^\infty \left\{ \frac{\partial}{\partial \tau}\left[K(t, \tau)\left(\frac{t}{\tau}\right)^\gamma\right] + \delta_0(\tau) \lim_{\tau \to +0} \left[\left(\frac{t}{\tau}\right)^\gamma K(t, \tau)\right] \right\} u(\tau, x)d\tau$$

$$= -t^\gamma \lim_{\tau \to \infty} [K(t, \tau)\tau^{-\gamma}u(\tau, x)] + t^\gamma \lim_{\tau \to 0+} [K(t, \tau)\tau^{-\gamma}u(\tau, x)]$$

$$+ \int_0^\infty K(t, \tau)\left(\frac{t}{\tau}\right)^\gamma \frac{\partial u(\tau, x)}{\partial \tau}d\tau - \lim_{\tau \to +0}\left[\left(\frac{t}{\tau}\right)^\gamma K(t, \tau)\right]u(0, x). \tag{7.96}$$

The first term on the right of (7.96) is zero by hypothesis *ii)* of the theorem. The sum of the second and last terms, which exist by hypothesis *i)*, also equals zero. Now taking equation (7.82) into account,

$$H_t v(t, x) = t^\gamma B(x, D_x) \int_0^\infty K(t, \tau)\tau^{-\gamma}u(\tau, x)d\tau + \frac{\gamma + 1}{2}t^\gamma A(x, D_x)v(t, x).$$

Further, since $E_0 = 0$, it follows that

$$\lim_{t \to 0} v(t, x) = \int_0^\infty \delta_0(\tau)u(\tau, x)d\tau = u(0, x),$$

which completes the proof. □

Let Π_γ denote the operator of multiplication by t^γ, i.e. $\Pi_\gamma h(t) = t^\gamma h(t)$, $h \in C(0, \infty)$. Applying Theorem 7.10 to the case $K(t, \tau) = f_{E_t^\beta}(\tau)$ in conjunction with Theorem 7.8, we obtain the following corollary.

Corollary 7.11. *Let* $\gamma \in (-1, 0]$ *and* $K(t, \tau) = f_{E_t^\beta}(\tau)$, *where* $f_{E_t^\beta}(\tau)$ *is defined in* (7.69). *Then* (i) $G_{-\gamma,t} = G_{\gamma,t}^{-1}$; (ii) $H_t = \Pi_\gamma G_{-\gamma,t} D_*^\beta$.

This Corollary yields an equivalent form for FPK equation (7.84) in the case when E_t^β is the inverse to the stable subordinator with index β and $\gamma \in (-1, 0]$:

$$H_t v(t, x) = t^\gamma G_{-\gamma,t} B(x, D_x)v(t, x) + \frac{\gamma + 1}{2}t^\gamma A(x, D_x)v(t, x), \tag{7.97}$$

with H_t as in Corollary 7.11.

Suppose the operator in the drift term $B(x, D_x) = 0$. Then equation (7.97) takes the form

$$H_t v(t, x) = \frac{\gamma + 1}{2} t^\gamma A(x, D_x) v(t, x). \tag{7.98}$$

Consequently, given an FPK equation associated to a non-time-changed fBM, the FPK equation for the corresponding time-changed fBM cannot be of the form: retain the right hand side and change the left hand side to a fractional derivative. Moreover, if a fractional derivative is desired on the left hand side in the time-changed case, then (7.84) shows that the right hand side must be a different operator from that in the non-time-changed case.

Notice that FPK equation (7.97) is valid for $\gamma \in (0, 1)$ as well. Indeed, part (ii) of Corollary 7.11 can be rewritten in the form $G_{\gamma,t} = G_{-\gamma,t}^{-1}$ for $\gamma > 0$. For $\gamma < 0$ part (ii) of Corollary 7.11 also implies $(G_{\gamma,t}^{-1})^{-1} = G_{-\gamma,t}^{-1} = G_{\gamma,t}$. Now applying operators $G_{-\gamma,t}$ and Π_γ consecutively to both sides of (7.84) we obtain (7.97) for all $\gamma \in (-1, 1)$.

Analogously, the FPK equation obtained in Theorem 7.9 with the mixing measure μ can be represented in its equivalent form as

$$H_t^\mu v(t, x) = t^\gamma G_{-\gamma,t}^\mu B(x, D_x) v(t, x)$$
$$+ \frac{\gamma + 1}{2} t^\gamma A(x, D_x) v(t, x), \quad t > 0, \ x \in \mathbb{R}^n, \tag{7.99}$$

where $H_t^\mu = \Pi_\gamma G_{\gamma,t}^\mu D_\mu$. We leave verification of the details to the reader.

The equivalence of equations (7.84) and (7.97) and the equivalence of equations (7.90) and (7.99) are obtained by means of Theorem 7.10. This fact can also be established with the help of the semigroup property of the family of operators $\{G_\gamma, -1 < \gamma < 1\}$:

$$G_\gamma g(t) = \int_0^\infty f_{E_t^\beta}(\tau) \tau^\gamma h(\tau) d\tau =: \mathcal{F}_\gamma h(t), \tag{7.100}$$

where $h \in C^\infty(0, \infty)$ is a non-negative bounded function. Denote the class of such functions by $\mathbf{U}$. Functions g and h in (7.100) are connected through the relation

$$g(t) = \int_0^\infty f_{E_t^\beta}(\tau) h(\tau) d\tau = \mathcal{F}_0 h(t) =: \mathcal{F} h(t). \tag{7.101}$$

It follows from the behavior of $f_{E_t^\beta}(\tau)$ as a function of t that $g \in C^\infty(0, \infty)$. On the other hand, obviously, operator $\mathcal{F}$ is bounded, $\|\mathcal{F} h\| \leqslant \|h\|$ in the sup-norm, and one-to-one due to positivity of $f_{E_t^\beta}(\tau)$. Therefore, the inverse $\mathcal{F}^{-1} : \mathcal{F} \mathbf{U} \to \mathbf{U}$ exists. Let a distribution $H(t, \tau)$ with $\operatorname{supp}(H) \subset (0, \infty) \times (0, \infty)$ be such that $\mathcal{F}^{-1} g(t) = \int_0^\infty H(t, \tau) g(\tau) d\tau$. Since $f_{E_t^\beta}(\tau) \in \mathcal{F} \mathbf{U}$ as a function of t for each $\tau > 0$, for an arbitrary $h \in \mathbf{U}$ one has

$$
\begin{aligned}
h(t) = \mathcal{F}^{-1}\mathcal{F}h(t) &= \int_0^\infty H(t,s)\left(\int_0^\infty f_{E_s^\beta}(\tau)h(\tau)d\tau\right)ds \\
&= \int_0^\infty h(\tau)\left(\int_0^\infty H(t,s)f_{E_s^\beta}(\tau)ds\right)d\tau \\
&= \left\langle \int_0^\infty H(t,s)f_{E_s^\beta}(\tau)ds, h(\tau)\right\rangle.
\end{aligned}
$$

We write this relation between $H(t,\tau)$ and $f_{E_t^\beta}(\tau)$ in the form

$$
\int_0^\infty H(t,s)f_{E_s^\beta}(\tau)ds = \delta_t(\tau). \tag{7.102}
$$

Proposition 7.1. *Let $-1 < \gamma < 1$, $-1 < \alpha < 1$, and $-1 < \gamma+\alpha < 1$. Then $G_\gamma \circ G_\alpha = G_{\gamma+\alpha}$.*

Proof. The proof uses the following two relations:

(1) $G_\gamma g(t) = \int_0^\infty \mathcal{F}_{\gamma,t}H(t,s)g(s)ds$, $\gamma \in (-1,1)$;

(2) $\int_0^\infty \mathcal{F}_{\gamma,t}H(t,s)\mathcal{F}_{\alpha,s}H(s,\tau)ds = \mathcal{F}_{\gamma+\alpha,t}H(t,\tau)$, with $-1 < \gamma, \alpha < 1$, and $-1 < \gamma+\alpha < 1$.

Indeed, using (7.100) and changing the order of integration, we obtain the first relation

$$
\begin{aligned}
G_\gamma g(t) &= \int_0^\infty f_{E_t^\beta}(\tau)\tau^\gamma\left(\int_0^\infty H(\tau,s)g(s)ds\right)d\tau \\
&= \int_0^\infty g(s)\left(\int_0^\infty f_{E_t^\beta}(\tau)H(\tau,s)\tau^\gamma d\tau\right)ds \\
&= \int_0^\infty \mathcal{F}_{\gamma,t}H(t,s)g(s)ds. \tag{7.103}
\end{aligned}
$$

It is readily seen that the internal integral in the second line of (7.103) is meaningful since $f_{E_t^\beta}(\tau)$ is a function of exponential decay when $\tau \to \infty$, which follows from (5.27). Further, in order to show the second relation, observe that

$$
\begin{aligned}
\int_0^\infty \mathcal{F}_{\gamma,t}H(t,s)\mathcal{F}_{\alpha,s}H(s,\tau)ds &= \int_0^\infty\left(\int_0^\infty f_{E_t^\beta}(p)H(p,s)p^\gamma dp\right)\left(\int_0^\infty f_{E_s^\beta}(q)H(q,\tau)q^\alpha dq\right)ds \\
&= \int_0^\infty\int_0^\infty f_{E_t^\beta}(p)H(q,\tau)p^\gamma q^\alpha\left(\int_0^\infty H(p,s)f_{E_s^\beta}(q)ds\right)dp\,dq.
\end{aligned}
$$

Due to (7.102), this equals

$$
\int_0^\infty f_{E_t^\beta}(p)p^\gamma\left(\int_0^\infty H(q,\tau)q^\alpha\delta_p(q)dq\right)dp = \int_0^\infty H(p,\tau)p^\alpha f_{E_t^\beta}(p)p^\gamma dp = \mathcal{F}_{\gamma+\alpha,t}H(t,\tau).
$$

Now we are ready to prove the claimed semigroup property. Making use of the two proved relations,

$$
\begin{aligned}
(G_\gamma \circ G_\alpha)g(t) &= G_\gamma\big[G_\alpha g(t)\big] \\
&= G_\gamma\Big[\int_0^\infty \mathcal{F}_{\alpha,t}H(t,s)g(s)ds\Big] \\
&= \int_0^\infty \mathcal{F}_{\gamma,t}H(t,s)\Big[\int_0^\infty \mathcal{F}_{\alpha,s}H(s,\tau)g(\tau)d\tau\Big]ds \\
&= \int_0^\infty g(\tau)\int_0^\infty \mathcal{F}_{\gamma,t}H(t,s)\mathcal{F}_{\alpha,s}H(s,\tau)ds\,d\tau \\
&= \int_0^\infty \mathcal{F}_{\gamma+\alpha,t}H(t,\tau)g(\tau)d\tau = G_{\gamma+\alpha}g(t),
\end{aligned}
$$

which completes the proof.   □

Remark 7.7.

(a) Proposition 7.1 immediately implies that $G_\gamma^{-1} = G_{-\gamma}$ for arbitrary $\gamma \in (-1,1)$. Indeed, $G_\gamma \circ G_{-\gamma} = G_0 = I$, as well as $G_{-\gamma} \circ G_\gamma = I$, where I is the identity operator. Thus, the statement in Corollary 7.11 is valid for all $\gamma \in (-1,1)$.

(b) Proposition 7.1 remains valid for the family $\{G_\gamma^\mu, -1 < \gamma < 1\}$ as well.

(c) As in [Meerschaert et al. (2009a)], if the governing equation for fBM B^H with Hurst parameter $H \in (0,1)$ is given by

$$
\frac{\partial h}{\partial t}(t,x) = 2Ht^{2H-1}a\,\frac{\partial^2 h}{\partial x^2}(t,x),
$$

then Theorem 7.8 and Proposition 7.1 imply that the governing equation for the time-changed fBM is either of the following equivalent forms:

$$
D_*^\beta h(t,x) = 2HG_{2H-1,t}\,a\,\frac{\partial^2 h}{\partial x^2}(t,x),
$$

$$
G_{1-2H,t}\,D_*^\beta h(t,x) = 2Ha\,\frac{\partial^2 h}{\partial x^2}(t,x).
$$

The method used in this section allows extension of results of Theorems 7.8–7.10 to the case of time-changed linear fractional stable motions. See [Meerschaert et al. (2009a)] for CTRW limits of correlated random variables whose limiting processes are time-changed fractional Brownian or linear fractional stable motions.

(d) The formula $v(t,x) = \mathcal{F}u(t,x)$ for a solution of FPK equations associated with time-changed fBM provides a useful tool for analysis of properties of a solution to initial value problems (7.84)–(7.85), (7.90)–(7.91), and (7.94)–(7.95).

(e) It is not necessary for the dependence of coefficients in (7.81) on t to be of the form t^γ. This function can be replaced by $[\nu(t)]^\gamma$, where $\nu(t)$ is a continuous function defined on $[0,\infty)$; however, the results essentially depend on the behavior of $\nu(t)$ near zero and infinity.

7.4.2 Applications

Two fractional generalizations of Black–Scholes models. Consider an application of fractional FPK equations to financial mathematics. In 1973, Fischer Black and Myron Scholes in [Black and Scholes (1973)] and Robert Merton in [Merton (1973)] suggested a stochastic model of a financial market in which a pricing formula for European call options is established. The model is usually referred to as the Black–Scholes (BS) model. The SDE representing the price of the underlying risky asset for an option and the deterministic equation which describes the change of the value of the option over time (called the Black–Scholes partial differential equation (BS PDE)) now form the classic theory of valuation of various types of options (contracts). For details of derivation and terminology of this theory, we refer the reader e.g. to books [Schoutens (2003), Hirsa and Neftci (2000), Steele (2001)]. The BS PDE takes the form

$$\frac{\partial u}{\partial t} = -\frac{\sigma^2 x^2}{2}\frac{\partial^2 u}{\partial x^2} - rx\frac{\partial u}{\partial x} + ru, \quad 0 < t < T, \ x \in \mathbb{R}_+, \tag{7.104}$$

where $u(t, x)$ is the option price at time t with the price x of the underlying risky asset (stock), σ is the volatility, r is the rate of interest, and T is the maturity time. Boundary conditions depend on the type of option and market. For instance, in the case of a European call option (in which the option can be exercised only on the maturity/expiration date), the terminal condition at $t = T$ is given by

$$u(T, x) = \phi(x), \quad x \in \mathbb{R}_+, \tag{7.105}$$

where $\phi(x) = \max\{x - K, 0\}$ with K being the price at which the call option can be exercised, called the strike price. Moreover, it is obvious that the option has no value (that is $u(t, x) = 0$) if the underlying stock has no value (that is $x = 0$) at any time. Similarly, $u(t, x) \sim x$ if x is large, leading to the following boundary conditions:

$$u(t, 0) = 0, \quad \text{and} \quad u(t, x) \sim x \text{ if } x \to \infty \text{ for all } 0 \leqslant t \leqslant T. \tag{7.106}$$

The price of the underlying stock is modeled by the SDE

$$dY_t = \mu Y_t dt + \sigma Y_t dB_t, \quad Y_0 = x_0, \tag{7.107}$$

where μ is the average rate of growth of the stock price, x_0 is the stock price at time $t = 0$, and σ is as above. It is not hard to verify using the Itô formula that the solution of SDE (7.107) is given by

$$Y_t = x_0 e^{(\mu - \frac{\sigma^2}{2})t + \sigma B_t}. \tag{7.108}$$

It is well known that the above-described BS model does not reflect a number of characteristic features present in forming real pricing processes. Such features include long-range dependence, power law decay of marginal distributions, skewness of distributions, memory effects, etc. Therefore, a number of generalizations of the BS model have been developed to reflect the above specific features. We refer the reader to the works [Hu and Øksendal (1999),Wyss (2000),Cheridito (2003),Sottinen and Valkeila (2003),Biagini et al. (2008),Magdziarz (2009c),Wang (2010),Wang, et. al. (2010)] and references therein for details. Below we will consider two different fractional generalizations of the BS model, which can be obtained from the results established in the previous sections of the current chapter.

Generalization 1: The first fractional generalization of the BS model is based on the time change of the driving process in (7.107). Notice that the change of variables $t = T - \tau$ in equation (7.104) reduces this equation to the form

$$\frac{\partial U(\tau, x)}{\partial \tau} = \frac{\sigma^2 x^2}{2} \frac{\partial^2 U(\tau, x)}{\partial x^2} + rx \frac{\partial U(\tau, x)}{\partial x} - rU(\tau, x), \tag{7.109}$$

$$0 < \tau < T, \ x \in \mathbb{R}_+,$$

and the condition (7.105) to the initial condition

$$U(0, x) = \phi(x), \quad x \in \mathbb{R}_+. \tag{7.110}$$

Here $U(\tau, x) \equiv u(T - \tau, x) = u(t, x)$. Let $E(\rho, \tau)$ be the inverse to a subordinator $U(\rho, \tau) \in \mathcal{S}$ with a mixing measure ρ; i.e. $U(\rho, \tau)$ has the Laplace transform $\mathbb{E}[e^{-sU(\rho,\tau)}] = e^{-\tau\Phi_\rho(s)}$ with $\Phi_\rho(s) = \int_0^1 s^\beta d\rho(\beta)$ for some finite positive measure ρ on $(0, 1)$. Then due to Theorem 7.4, the time-changed process

$$X_\tau = Y_{E(\rho,\tau)} = x_0 e^{(\mu - \frac{\sigma^2}{2})E(\rho,\tau) + \sigma B_{E(\rho,\tau)}}$$

satisfies SDE

$$dX_\tau = \mu X_\tau dE(\rho, \tau) + \sigma X_\tau dB_{E(\rho,\tau)}, \quad Y_0 = x_0. \tag{7.111}$$

Moreover, the function $V(\tau, x) = \int_0^\infty f_{E(\rho,\tau)}(s)U(s, x)ds$ satisfies a time-distributed fractional order differential equation of the form

$$D_\rho V(\tau, x) = \frac{\sigma^2 x^2}{2} \frac{\partial^2 V(\tau, x)}{\partial x^2} + rx \frac{\partial V(\tau, x)}{\partial x} - rV(\tau, x), \tag{7.112}$$

$$0 < \tau < T, \ x \in \mathbb{R}_+,$$

with the same initial condition as (7.110), that is,

$$V(0, x) = \phi(x), \quad x \in \mathbb{R}_+.$$

If $\rho(du) = \delta_\beta(u)du$, that is, $E(\rho, \tau) = E_\tau^\beta$ is the inverse to a stable subordinator with index β, then equation (7.112) reduces to

$$D_*^\beta V(\tau, x) = \frac{\sigma^2 x^2}{2} \frac{\partial^2 V(\tau, x)}{\partial x^2} + rx \frac{\partial V(\tau, x)}{\partial x} - rV(\tau, x), \tag{7.113}$$

$$0 < \tau < T, \ x \in \mathbb{R}_+,$$

where D_*^β is the fractional derivative of order β in the sense of Caputo–Djrbashian.

A numerical scheme for equation (7.113) with the Riemann–Liouville fractional derivative in place of D_*^β is considered in [Song and Wang (2013)]. In [Magdziarz (2009c)], an analogue of the classical BS formula with the driving process being the time-changed Brownian motion $B_{E_t^\beta}$ is established, where some implications of the fractional BS model for financial markets are also studied. Namely, the paper shows that the market corresponding to the BS model given by (7.111) is *arbitrage-free*, but it is *incomplete*. It is known that the market model defined by a triple $(\Omega, \mathcal{F}, \mathbb{P})$ and an asset price process $\{Y_t : t \in [0, T]\}$ is arbitrage-free if and only if there exists a probability measure $\mathbb{Q}$ equivalent to $\mathbb{P}$ such that the process $\{Y_t : t \in [0, T]\}$ is a martingale with respect to the measure $\mathbb{Q}$. In an arbitrage-free market, there is no way to make a risk-free profit. On the other hand, a market modeled by the asset price process $\{Y_t : t \in [0, T]\}$ is *complete* if and only if there is a unique martingale measure $\mathbb{Q}$ equivalent to $\mathbb{P}$. Otherwise, the market is called *incomplete*. A generalization of the BS model to the case when the driving process in (7.107) is replaced by an α-stable Lévy process appears in e.g. [Cartea and Del-Castillo-Negrete (2007)].

Generalization 2: The second fractional generalization of the BS model is based on SDE (7.107) with the driving process replaced by fBM B_t^H with Hurst parameter $H \in (1/2, 1)$ (see e.g. [Rogers (1997), Hu and Øksendal (1999), Cheridito (2003), Sottinen and Valkeila (2003), Osu and Ifeoma (2016)]). Namely, the SDE considered for the underlying stock price process is given by

$$dY_t = \mu Y_t dt + \sigma Y_t dB_t^H, \quad Y_0 = x_0. \tag{7.114}$$

However, since there are several approaches to defining the stochastic integral driven by fBM, different definitions may lead to models of differing natures (to be discussed in Remark 7.8 below). In particular, the explicit form of the solution Y_t may differ depending on the definition (see e.g. [Sottinen and Valkeila (2003)]). Paper [Osu and Ifeoma (2016)] obtains a BS PDE connected with the driving process B_t^H in the form

$$\frac{\partial u}{\partial t} = -Ht^{2H-1}\sigma^2 x^2 \frac{\partial^2 u}{\partial x^2} - rx\frac{\partial u}{\partial x} + ru, \quad 0 < t < T, \ x \in \mathbb{R}_+. \tag{7.115}$$

Setting $2H - 1 = \gamma$, and

$$B\varphi(x) = -rx\frac{\partial \varphi(x)}{\partial x} + r\varphi(x), \quad \text{and} \quad A\varphi(x) = -\sigma^2 x^2 \frac{\partial^2 \varphi(x)}{\partial x^2},$$

with the

$$\text{Dom}(A) = \{\varphi \in C^2[0, \infty) : \varphi(0) = 0, \ \varphi(x) \sim x, \ x \to \infty\},$$

one can rewrite (7.115) as

$$\frac{\partial u(t, x)}{\partial t} = Bu(t, x) + \frac{\gamma + 1}{2}t^\gamma Au(t, x), \quad 0 < t < T. \tag{7.116}$$

Thus, the BS PDE in the case when the driving process is fBM can be written in the form (7.82), and therefore, Theorems 7.8 and 7.9 are applicable.

Let $E(\rho, t)$ be the inverse to $U(\rho, t) \in \mathcal{S}$. By Theorem 7.9, the function $V(t, x) = \int_0^\infty f_{E(\rho,t)}(s)u(s, x)ds$ satisfies a fractional version of the BS PDE

$$D_\rho V(t, x) = BV(t, x) + \frac{\gamma + 1}{2}G_{\gamma,t}^\rho AV(t, x), \quad 0 < t < T, \tag{7.117}$$

where the operator $G_{\gamma,t}^\rho$ is defined in (7.92). The latter equation in the explicit form through the Hurst parameter takes the form

$$D_\rho V(t,x) = -H\sigma^2 x^2 G_{2H-1,t}^\rho \frac{\partial^2 V(t,x)}{\partial x^2}$$

$$- rx\frac{\partial V(t,x)}{\partial x} + rV(t,x), \quad 0 < t < T, \ x \in \mathbb{R}_+. \tag{7.118}$$

Using the operator $H_t^\rho = \Pi_\gamma G_{\gamma,t}^\rho D_\rho$ with $\gamma = 2H - 1$ (see equation (7.99)), we can also represent equation (7.118) in the form

$$H_t^\rho V(t,x) = t^{2H-1} G_{1-2H,t}^\rho \left[-rx\frac{\partial V(t,x)}{\partial x} + rV(t,x) \right]$$

$$- Ht^{2H-1}\sigma^2 x^2 \frac{\partial^2 V(t,x)}{\partial x^2}, \quad t > 0, \ x \in \mathbb{R}_+. \tag{7.119}$$

In the case when the measure ρ is given by $\rho(du) = \delta_\beta(u)du$, that is, $E(\rho,t) = E_t^\beta$ is the time-change inverse to a stable subordinator with index $\beta \in (0,1)$, equation (7.118) takes the form

$$D_*^\beta V(t,x) = -H\sigma^2 x^2 G_{2H-1,t} \frac{\partial^2 V(t,x)}{\partial x^2}$$

$$- rx\frac{\partial V(t,x)}{\partial x} + rV(t,x), \quad 0 < t < T, \ x \in \mathbb{R}_+, \tag{7.120}$$

where D_*^β is the fractional derivative of order β in the sense of Caputo–Djrbashian and $G_{2H-1,t}$ is the operator in (7.86). Similarly, in this particular case, equation (7.119) reduces to the form

$$H_t V(t,x) = t^{2H-1} G_{1-2H,t} \left[-rx\frac{\partial V(t,x)}{\partial x} + rV(t,x) \right]$$

$$- Ht^{2H-1}\sigma^2 x^2 \frac{\partial^2 V(t,x)}{\partial x^2}, \quad t > 0, \ x \in \mathbb{R}_+, \tag{7.121}$$

where the operator H_t is defined in (7.93).

Remark 7.8. The market based on the classical BS model in which the driving process is Brownian motion is arbitrage-free and complete. However, as noted above, the market is incomplete if the driving process is a time-changed Brownian motion, where the time-change is the inverse of a stable subordinator. On the other hand, if the driving process is fBM, whether the market admits arbitrage or not depends on how the stochastic integral is defined (see [Hu and Øksendal (1999)]). A challenging question is to study similar properties of markets modeled by more general asset pricing processes in which the driving process is a time-changed fBM with the time-change being the inverse to a subordinator belonging to the class $\mathcal{S}$.

7.5 FPK equations associated with general Gaussian processes

This brief section derives FPK equations for general Gaussian processes, which include fractional Brownian motion (fBM) discussed in Section 7.3 and Volterra processes introduced in Section 5.5. As we will see, the variance functions of the Gaussian processes play a key role in the derivation of the FPK equations. A result obtained in this section (Proposition 7.2) will be used in Section 7.6 to derive fractional FPK equations associated with time-changed Gaussian processes (Theorems 7.12 and 7.13). These results generalize the following correspondence between the two FPK equations associated with an n-dimensional fBM B_t^H with Hurst parameter $H \in (0,1)$ and its time-changed counterpart $B_{E_t^\beta}^H$ (see Remark 7.7 (c)):

$$\partial_t p(t,x) = Ht^{2H-1}\Delta p(t,x); \tag{7.122}$$

$$D_{*,t}^\beta q(t,x) = HG_{2H-1,t}^\beta \Delta q(t,x), \tag{7.123}$$

with $G_{2H-1,t}^\beta$ being the operator in (7.86) with $\gamma = 2H - 1$. Here, $\partial_t = \frac{\partial}{\partial t}$ and $\Delta = \sum_{j=1}^n \partial_{x^j}^2 = \sum_{j=1}^n \left(\frac{\partial}{\partial x^j}\right)^2$, with the vector $x \in \mathbb{R}^n$ denoted as $x = (x^1, \dots, x^n)$. As noted in Section 7.4, the right hand side of (7.123) has a different form than that of (7.122). Namely, the operator $G_{2H-1,t}^\beta$ instead of t^{2H-1} appears, which is ascribed to dependence between increments over non-overlapping intervals of the fBM. A similar correspondence will be observed between FPK equations for a general Gaussian process and its time-changed version.

Recall that a one-dimensional stochastic process $X = (X_t)_{t\geqslant 0}$ is called a *Gaussian process* if the random vector $(X_{t_1}, \dots, X_{t_m})$ has a multivariate Gaussian distribution for all finite sequences $0 \leqslant t_1 < \cdots < t_m < \infty$. The joint distributions are characterized by the mean function $\mathbb{E}[X_t]$ and the covariance function $R_X(s,t) = \text{Cov}(X_s, X_t)$. The covariance function of a given zero-mean Gaussian process is symmetric and positive semi-definite. Conversely, every symmetric, positive semi-definite function on $[0,\infty) \times [0,\infty)$ is the covariance function of some zero-mean Gaussian process (see e.g., Theorem 8.2 of [Janson (1997)]). Examples of such functions include $R_X(s,t) = s \wedge t$ for Brownian motion, $R_X(s,t) = R_H(s,t)$ in (7.57) for fBM with Hurst parameter $H \in (0,1)$, and $R_X(s,t) = \sigma_0^2 + s \cdot t$ which is obtained from linear regression (see [Rasmussen and Williams (2006)]). The sum and the product of two covariance functions for Gaussian processes are again covariance functions for some Gaussian processes. For more examples of covariance functions, consult e.g. [Rasmussen and Williams (2006)].

In Sections 7.5 and 7.6, we restrict our attention to n-dimensional zero-mean Gaussian processes $X = (X^1, \dots, X^n)$ *whose components X^j are assumed to be independent one-dimensional Gaussian processes starting at 0 with (possibly distinct) positive definite covariance functions $R_{X^j}(s,t)$. The variance functions are denoted by $R_{X^j}(t) = R_{X^j}(t,t) = \mathbb{E}[(X_t^j)^2]$. For differentiable variance functions $R_{X^j}(t)$, let*

$$A = \frac{1}{2}\sum_{j=1}^n R_{X^j}'(t)\partial_{x^j}^2. \tag{7.124}$$

Clearly $A = \frac{1}{2}\Delta$ if X is an n-dimensional Brownian motion.

Proposition 7.2. *Let $X = (X^1, \ldots, X^n)$ be an n-dimensional zero-mean Gaussian process with covariance functions $R_{X^j}(s,t)$, $j = 1, \ldots, n$. Suppose the variance functions $R_{X^j}(t) = R_{X^j}(t,t)$ are differentiable on $(0,\infty)$. Then the transition probabilities $p(t,x)$ of X satisfy the PDE*

$$\partial_t p(t,x) = Ap(t,x), \ t > 0, \ x \in \mathbb{R}^n, \tag{7.125}$$

with initial condition $p(0,x) = \delta_0(x)$, where A is the operator in (7.124) and $\delta_0(x)$ is the Dirac delta function with mass on 0.

Proof. Since the components X^j of X are assumed to be independent zero-mean Gaussian processes, it follows that

$$p(t,x) = \prod_{j=1}^{n} (2\pi R_{X^j}(t))^{-1/2} \times \exp\left\{ -\sum_{j=1}^{n} \frac{(x^j)^2}{2R_{X^j}(t)} \right\}. \tag{7.126}$$

Direct computation of partial derivatives of $p(t,x)$ yields the equality in (7.125). Moreover, the initial condition $p(0,x) = \delta_0(x)$ follows immediately from the assumption that the process X starts at 0. □

Remark 7.9. a) In Proposition 7.2, if the components X^j are independent Gaussian processes with a *common* variance function $R_X(t)$ which is differentiable, then (7.125) reduces to the following form:

$$\partial_t p(t,x) = \frac{1}{2} R'_X(t)\Delta p(t,x). \tag{7.127}$$

In particular, if $X = B^H$ is an n-dimensional fBM with Hurst parameter $H \in (0,1)$, then it follows from (7.57) that $R_X(t) = R_H(t,t) = t^{2H}$ and $R'_X(t) = 2Ht^{2H-1}$, which recovers the FPK equation (7.122).

b) If $X = (X^1, \ldots, X^n)$ is an n-dimensional Gaussian process with mean functions $m_{X^j}(t)$ and covariance functions $R_{X^j}(s,t)$, and if both $m_{X^j}(t)$ and $R_{X^j}(t) = R_{X^j}(t,t)$ are differentiable, then the associated FPK equation contains an additional term:

$$\partial_t p(t,x) = Ap(t,x) + Bp(t,x), \tag{7.128}$$

where

$$B = -\sum_{j=1}^{n} m'_{X^j}(t)\partial_{x^j}. \tag{7.129}$$

Such Gaussian processes include e.g. the process defined by the sum of a Brownian motion and a deterministic differentiable function.

c) PDE (7.125) is a parabolic equation due to the assumption that the covariance functions are positive definite. Therefore, uniqueness of the solution to PDE (7.125) with initial condition $p(0,x) = \varphi(x)$, where $\varphi(x)$ is an element of some function space, is guaranteed (see [Friedman (1964)]).

7.6 Fractional FPK equations for time-changed Gaussian processes

7.6.1 *Theory*

Theorems 7.12 and 7.13 formulate the FPK equation for a time-changed Gaussian process under the assumption that the time-change process is independent of the Gaussian process. As in Proposition 7.2, the variance function plays a key role here. Recall that we only consider zero-mean Gaussian processes starting at 0 which have independent components with positive definite covariance functions. As usual, J_t^α denotes the fractional integral operator of order α acting on t, while $\tilde{g}(s) = L[g](s) = L_{t \to s}[g(t)](s)$ and $L^{-1}[f](t) = L_{s \to t}^{-1}[f(s)](t)$ denote the Laplace transform and the inverse Laplace transform, respectively.

Theorem 7.12. *Let $X = (X^1, \ldots, X^n)$ be an n-dimensional zero-mean Gaussian process with covariance functions $R_{X^j}(s,t)$, $j = 1, \ldots, n$, and let E^β be the inverse of a stable subordinator U^β of index $\beta \in (0,1)$, independent of X. Suppose the variance functions $R_{X^j}(t) = R_{X^j}(t,t)$ are differentiable on $(0,\infty)$ and Laplace transformable. Then the transition probabilities $q(t,x)$ of the time-changed Gaussian process $(X_{E_t^\beta})$ satisfy the equivalent PDEs*

$$D_{*,t}^\beta q(t,x) = \sum_{j=1}^n J_t^{1-\beta} \Lambda_{X^j,t}^\beta \, \partial_{x^j}^2 q(t,x), \ t > 0, \ x \in \mathbb{R}^n, \tag{7.130}$$

$$\partial_t q(t,x) = \sum_{j=1}^n \Lambda_{X^j,t}^\beta \, \partial_{x^j}^2 q(t,x), \ t > 0, \ x \in \mathbb{R}^n, \tag{7.131}$$

with initial condition $q(0,x) = \delta_0(x)$. Here, $\Lambda_{X^j,t}^\beta$, $j = 1, \ldots, n$, are the operators acting on t given by

$$\Lambda_{X^j,t}^\beta g(t) = \frac{\beta}{2} L_{s \to t}^{-1} \left[\frac{1}{2\pi i} \int_C (s^\beta - z^\beta) \widetilde{R_{X^j}}(s^\beta - z^\beta) \tilde{g}(z) \, dz \right](t), \tag{7.132}$$

with $z^\beta = e^{\beta \mathrm{Ln}(z)}$, $\mathrm{Ln}(z)$ being the principal value of the complex logarithmic function $\ln(z)$ with cut along the negative real axis, and C being a curve in the complex plane obtained via the transformation $\zeta = z^\beta$ which leaves all the singularities of $\widetilde{R_{X^j}}$ on one side.

Proof. Let $p(t,x)$ denote the transition probabilities of the Gaussian process X. For each $x \in \mathbb{R}^n$, it follows from the independence assumption between E^β and X that

$$q(t,x) = \int_0^\infty f_{E_t^\beta}(\tau) p(\tau,x) \, d\tau, \ t > 0. \tag{7.133}$$

Relationship (7.133) and the equality (see part (d) of Lemma 7.1)

$$L_{t \to s}[f_{E_t^\beta}(\tau)](s) = s^{\beta-1} e^{-\tau s^\beta}$$

together yield

$$\tilde{q}(s,x) = s^{\beta-1} \tilde{p}(s^\beta, x), \ s > 0. \tag{7.134}$$

Since $R_{X^j}(t)$ is Laplace transformable, $\widetilde{R_{X^j}}(s)$ exists for all $s > a$ for some constant $a \geqslant 0$. Taking Laplace transforms on both sides of (7.125),

$$s\tilde{p}(s,x) - p(0,x) = \frac{1}{2}\sum_{j=1}^{n} L_{t\to s}[R'_{X^j}(t)\partial^2_{x^j}p(t,x)](s) \tag{7.135}$$

$$= \frac{1}{2}\sum_{j=1}^{n}\left(L_t[R'_{X^j}(t)] * L_t[\partial^2_{x^j}p(t,x)]\right)(s)$$

$$= \frac{1}{2}\sum_{j=1}^{n}\left[\frac{1}{2\pi i}\int_{c-i\infty}^{c+i\infty}\widetilde{R'_{X^j}}(s-\zeta)\partial^2_{x^j}\tilde{p}(\zeta,x)\,d\zeta\right]$$

$$= \frac{\beta}{2}\sum_{j=1}^{n}\left[\frac{1}{2\pi i}\int_C \widetilde{R'_{X^j}}(s-z^\beta)\partial^2_{x^j}\tilde{p}(z^\beta,x)z^{\beta-1}\,dz\right], \quad s > a,$$

where $*$ denotes the convolution of Laplace images and the function

$$\widetilde{R'_{X^j}}(s) = s\widetilde{R_{X^j}}(s), \quad s > a, \tag{7.136}$$

exists by assumption. Equation (7.136) is valid since $R_{X^j}(0) = 0$ due to the initial condition $X^j(0) = 0$. Moreover, since $E_0^\beta = 0$ with probability one, it follows that $q(0,x) = p(0,x) = \delta_0(x)$. Replacing s by s^β and using the identity (7.134) yields

$$s\tilde{q}(s,x) - q(0,x) = \frac{\beta}{2}\sum_{j=1}^{n}\left[\frac{1}{2\pi i}\int_C \widetilde{R'_{X^j}}(s^\beta - z^\beta)\partial^2_{x^j}\tilde{q}(z,x)\,dz\right], \quad s > a^{1/\beta}. \tag{7.137}$$

Since the left hand side equals $L_{t\to s}[\partial_t q(t,x)](s)$, PDE (7.131) follows upon substituting (7.136) and taking the inverse Laplace transform on both sides. Moreover, applying the fractional integral operator $J_t^{1-\beta}$ to both sides of (7.131) yields (7.130). □

Remark 7.10. a) Representation (7.133) yields the estimate

$$\sup_{t>0,x\in\mathbb{R}^n}|q(t,x)| \leqslant \sup_{t>0,x\in\mathbb{R}^n}|p(t,x)|,$$

which, together with the uniqueness of the solution to PDE (7.125) with $p(0,x) = \delta_0(x)$, guarantees uniqueness of PDE (7.130) (or PDE (7.131)) with initial condition $q(0,x) = \delta_0(x)$. The same argument applies to the PDEs to be established in Theorem 7.13 as well.

b) Representation (7.133) provides a bridge to approximation of the unique solution to the equivalent PDEs (7.130) and (7.131).

The next theorem extends the previous theorem to time-changes which are the inverses of mixtures of independent stable subordinators belonging to the class $\mathcal{S}$ introduced in Section 7.2. Recall that $U(\mu,t) \in \mathcal{S}$ means that $U(\mu,t)$ is a strictly increasing subordinator with Laplace transform

$$\mathbb{E}[e^{-sU(\mu,t)}] = e^{-t\Phi_\mu(s)} \quad \text{with} \quad \Phi_\mu(s) = \int_0^1 s^\beta\,d\mu(\beta), \tag{7.138}$$

where μ is a finite positive measure on $(0,1)$. As usual, D_μ denotes the Caputo–Djrbashian distributed order differential operator with the mixing measure μ:

$$D_\mu q(t,x) = \int_0^1 D_{*,t}^\beta q(t,x)\, d\mu(\beta).$$

Theorem 7.13. *Let $X = (X^1, \ldots, X^n)$ be an n-dimensional zero-mean Gaussian process with covariance functions $R_{X^j}(s,t)$, $j = 1, \ldots, n$, and let $E(\mu,t)$ be the inverse of a subordinator $U(\mu,t) \in \mathcal{S}$ independent of X. Suppose the variance functions $R_{X^j}(t) = R_{X^j}(t,t)$ are differentiable on $(0,\infty)$ and Laplace transformable. Then the transition probabilities $q(t,x)$ of the time-changed Gaussian process $(X_{E(\mu,t)})$ satisfy the PDEs*

$$D_\mu q(t,x) = \sum_{j=1}^n \int_0^1 J_t^{1-\beta} \Lambda_{X^j,t}^\mu \, \partial_{x^j}^2 q(t,x)\, d\mu(\beta), \quad t > 0,\ x \in \mathbb{R}^n, \tag{7.139}$$

and

$$\partial_t q(t,x) = \sum_{j=1}^n \Lambda_{X^j,t}^\mu \, \partial_{x^j}^2 q(t,x), \quad t > 0,\ x \in \mathbb{R}^n, \tag{7.140}$$

with initial condition $q(0,x) = \delta_0(x)$, where $\Lambda_{X^j,t}^\mu$, $j = 1, \ldots, n$, are the operators acting on t given by

$$\Lambda_{X^j,t}^\mu g(t) = \frac{1}{2} L_{s \to t}^{-1} \left[\frac{1}{2\pi i} \int_{\mathcal{C}} \left(\Phi_\mu(s) - \Phi_\mu(z) \right) \widetilde{R_{X^j}} \left(\Phi_\mu(s) - \Phi_\mu(z) \right) m_\mu(z) \tilde{g}(z)\, dz \right](t), \tag{7.141}$$

with

$$\Phi_\mu(z) = \int_0^1 e^{\beta \operatorname{Ln}(z)}\, d\mu(\beta) \quad \text{and} \quad m_\mu(z) = \frac{1}{\Phi_\mu(z)} \int_0^1 \beta z^\beta\, d\mu(\beta),$$

and $\mathcal{C}$ being a curve in the complex plane obtained via the transformation $\zeta = \Phi_\mu(z)$ which leaves all the singularities of $\widetilde{R_{X^j}}$ on one side.

Proof. We only sketch the proof since it is similar to the proof of Theorem 7.12. Let $p(t,x)$ denote the transition probabilities of the Gaussian process X. For each $x \in \mathbb{R}^n$, it follows from relationship (7.133) with $f_{E_t^\beta}$ replaced by $f_{E(\mu,t)}$ together with part (d) of Lemma 7.3 that

$$\tilde{q}(s,x) = \frac{\Phi_\mu(s)}{s} \tilde{p}(\Phi_\mu(s),x), \quad s > 0. \tag{7.142}$$

Taking Laplace transforms on both sides of (7.125) leads to the second to last equality in (7.135). Letting $\zeta = \Phi_\mu(z)$ yields

$$s\tilde{p}(s,x) - p(0,x) = \frac{1}{2} \sum_{j=1}^n \left[\frac{1}{2\pi i} \int_{\mathcal{C}} \widetilde{R_{X^j}'} (s - \Phi_\mu(z)) \partial_{x^j}^2 \tilde{p}(\Phi_\mu(z),x) \frac{\Phi_\mu(z)}{z} m_\mu(z)\, dz \right],$$

which is valid for all s for which $\widetilde{R_{X^j}}(s)$ exists. Replacing s by $\Phi_\mu(s)$ and using the identity (7.142) yields an equation similar to (7.137). PDE (7.140) is obtained upon taking the inverse Laplace transform on both sides. Finally, applying the fractional integral operator $J_t^{1-\beta}$ and integrating with respect to μ on both sides of (7.140) yields (7.139). $\qquad\square$

Remark 7.11. a) If $\mu = \delta_{\beta_0}$ is a Dirac measure with mass at $\beta_0 \in (0,1)$, then $\Lambda^{\mu}_{X^j,t} g(t) = \Lambda^{\beta_0}_{X^j,t} g(t)$ and the FPK equations in (7.139) and (7.140) respectively reduce to the FPK equations in (7.130) and (7.131) with $\beta = \beta_0$, as expected.

b) In Theorem 7.13, if the components X^j are independent Gaussian processes with a *common* variance function $R_X(t)$ which is differentiable and Laplace transformable, then the FPK equations in (7.139) and (7.140) respectively reduce to the following simple forms:

$$D_\mu q(t,x) = \int_0^1 J_t^{1-\beta} \Lambda^{\mu}_{X,t} \Delta q(t,x) \, d\mu(\beta);$$ (7.143)

$$\partial_t q(t,x) = \Lambda^{\mu}_{X,t} \Delta q(t,x).$$ (7.144)

7.6.2   Applications

This subsection is devoted to applications of Proposition 7.2 and Theorems 7.12 and 7.13 concerning FPK equations for Gaussian and time-changed Gaussian processes. For simplicity of discussion, we will consider the time-change E^β given by the inverse of a single stable subordinator U^β of index $\beta \in (0,1)$, rather than the more general time-change process $E(\mu,t)$.

Application 1 (Fractional Brownian motion). Let B^H be an n-dimensional fractional Brownian motion (fBM) and let E^β be the inverse of a stable subordinator of index $\beta \in (0,1)$, independent of B^H. Then the components of B^H share the common variance function $R_{B^H}(t) = t^{2H}$ and its Laplace transform $\widetilde{R'_{B^H}}(s) = 2H\Gamma(2H)/s^{2H}$. Hence, Proposition 7.2 and Theorem 7.12 immediately recover the FPK equations in (7.122) and (7.123) for the fBM B^H and the time-changed fBM $(B^H_{E^\beta_t})$. In this case,

$$J_t^{1-\beta} \Lambda^\beta_{B^H,t} = HG^\beta_{2H-1,t},$$ (7.145)

where $G^\beta_{\gamma,t}$ is the operator given in (7.86). Note that the curve $\mathcal{C}$ appearing in the expression of the operator $\Lambda^\beta_{B^H,t}$ in (7.132) can be replaced by a vertical line $\{C + ir; r \in \mathbb{R}\}$ with $0 < C < s$ since the integrand has a singularity only at $z = s$.

Application 2 (Mixed fractional Brownian motion). Let $X = (X_t)_{t \geq 0}$ be an n-dimensional process defined by a finite linear combination of independent zero-mean Gaussian processes $X_1, \ldots, X_m$:

$$X_t = \sum_{\ell=1}^m a_\ell X_{\ell,t}$$

with $a_1, \ldots, a_m \in \mathbb{R}$. For simplicity, assume that for each $\ell = 1, \ldots, m$, the components of the vector $X_\ell = (X_\ell^1, \ldots, X_\ell^n)$ share a common variance function $R_{X_\ell}(t)$. Then the process X is again a Gaussian process whose components have the same variance function $R_X(t) = \sum_{\ell=1}^m a_\ell^2 R_{X_\ell}(t)$. Therefore, it follows from Proposition

7.2 and Theorem 7.12 that the FPK equations for X and $(X_{E_t^\beta})$, under the independence assumption between E^β and X, are respectively given by

$$\partial_t p(t, x) = \phi(t) \Delta p(t, x) \quad \text{and} \quad D^\beta_{*,t} q(t, x) = \Phi^\beta_t \Delta q(t, x), \tag{7.146}$$

where $\phi(t) = \frac{1}{2} \sum_{\ell=1}^m a_\ell^2 R'_{X_\ell}(t)$ and $\Phi^\beta_t = \sum_{\ell=1}^m a_\ell^2 J_t^{1-\beta} \Lambda^\beta_{X_\ell,t}$. Notice that $\phi(t)$ simply denotes the multiplication by a function of t whereas Φ^β_t is an operator acting on t. This generalizes the correspondence between the function t^{2H-1} and the operator $G^\beta_{2H-1,t}$ observed in the FPK equations in (7.122) and (7.123) for the fBM and the time-changed fBM.

A *mixed fractional Brownian motion* is a finite linear combination of independent fBMs (see [Miao et al. (2008), Thäle (2009)] for its properties). It was introduced in [Cheridito (2001)] to discuss the price of a European call option on an asset driven by the process. The process X considered in that paper is of the form $X_t = B_t + aB_t^H$, where $a \in \mathbb{R}$, B is a Brownian motion, and B^H is an fBM with Hurst parameter $H \in (0, 1)$. In this situation, the FPK equations in (7.146), with the help of (7.145), yield

$$\partial_t p(t, x) = \frac{1}{2} \Delta p(t, x) + a^2 H t^{2H-1} \Delta p(t, x); \tag{7.147}$$

$$D^\beta_{*,t} q(t, x) = \frac{1}{2} \Delta q(t, x) + a^2 H G^\beta_{2H-1,t} \Delta q(t, x). \tag{7.148}$$

Application 3 (Fractional Brownian motion with variable Hurst parameter). *Volterra processes* form an important subclass of Gaussian processes. They are continuous zero-mean Gaussian processes $V = (V_t)$ defined on a given finite interval $[0, T]$ with integral representations of the form $V_t = \int_0^t K(t, s) dB_s$ for some deterministic kernel $K(t, s)$ and Brownian motion B (see Section 5.5 for details). As noted in Section 5.5, an fBM is clearly an example of a Volterra process. In particular, an fBM B^H with Hurst parameter $H \in (1/2, 1)$ is represented as $B_t^H = \int_0^t K_H(t, s) dB_s$ with the kernel

$$K_H(t, s) = c_H s^{1/2-H} \int_s^t (r - s)^{H-3/2} r^{H-1/2} dr, \quad t > s, \tag{7.149}$$

where the positive constant c_H is chosen so that the integral $\int_0^{t \wedge s} K_H(t, r) K_H(s, r) dr$ coincides with the covariance function $R_H(s, t)$ in (7.57). Increments of B^H exhibit long range dependence.

A particular interesting Volterra process is the fBM with time-dependent Hurst parameter $H(t)$ suggested in Theorem 9 of [Decreusefond (2005)]. Namely, suppose $H(t) : [0, T] \to (1/2, 1)$ is a deterministic function satisfying the following conditions:

$$\inf_{t \in [0,T]} H(t) > \frac{1}{2} \quad \text{and} \quad H(t) \in \mathcal{S}_{1/2+\alpha,2} \quad \text{for some} \quad \alpha \in \left(0, \inf_{t \in [0,T]} H(t) - \frac{1}{2}\right), \tag{7.150}$$

where $\mathcal{S}_{\eta,2}$ is the Sobolev–Slobodetzki space given by the closure of the space $C^1[0,T]$ with respect to the semi-norm

$$\|f\|^2 = \int_0^T \int_0^T \frac{|f(t) - f(s)|^2}{|t-s|^{1+2\eta}} \, dt \, ds. \tag{7.151}$$

Then representation (7.149) with H replaced by $H(t)$ induces a covariance function $R_V(s,t) = \int_0^{t \wedge s} K_{H(t)}(t,r) K_{H(s)}(s,r) \, dr$ for some Volterra process V on $[0,T]$. The variance function is given by $R_V(t) = t^{2H(t)}$ and is necessarily continuous due to the Sobolev embedding theorem, which says $\mathcal{S}_{\eta,2} \subset C[0,T]$ for all $\eta > 1/2$ (see e.g. [Hörmander (2007)]). Therefore, $H(t)$ is also continuous.

Let $H(t) : [0,\infty) \to (1/2, 1)$ be a differentiable function whose restriction to any finite interval $[0,T]$ satisfies the conditions in (7.150). For each $T > 0$, let $K_{V^T}(s,t)$ be the kernel inducing the covariance function $R_{V^T}(s,t)$ of the associated Volterra process V^T defined on $[0,T]$ as above. The definition of $K_{V^T}(s,t)$ is consistent; i.e. $K_{V^{T_1}}(s,t) = K_{V^{T_2}}(s,t)$ for any $0 \leqslant s, t \leqslant T_1 \leqslant T_2 < \infty$. Hence, so is that of $R_{V^T}(s,t)$, which implies that the function $R_X(s,t)$ given by $R_X(s,t) = R_{V^T}(s,t)$ whenever $0 \leqslant s, t \leqslant T < \infty$ is a well-defined covariance function of a Gaussian process X on $[0,\infty)$ whose restriction to each interval $[0,T]$ coincides with V^T. The process X represents an fBM with variable Hurst parameter. The variance function $R_X(t) = t^{2H(t)}$ is differentiable on $(0,\infty)$ by assumption and Laplace transformable due to the estimate $R_X(t) \leqslant t^2$. Therefore, Proposition 7.2 and Theorem 7.12 can be applied to yield the FPK equations for X and the time-changed process $(X_{E_t^\beta})$ under the independence assumption between E^β and X.

Application 4 (Fractional Brownian motion with piecewise constant Hurst parameter). The fBM discussed in the previous example has a continuously varying Hurst parameter $H(t) : [0,\infty) \to (1/2, 1)$. Here we consider a piecewise constant Hurst parameter $H(t) : [0,\infty) \to (0,1)$ which is described as

$$H(t) = \sum_{k=0}^N H_k \boldsymbol{I}_{[T_k, T_{k+1})}(t), \tag{7.152}$$

where $\{H_k\}_{k=0}^N$ are constants in $(0,1)$, $\{T_k\}_{k=0}^N$ are fixed times such that $0 = T_0 < T_1 < \cdots < T_N < T_{N+1} = \infty$, and $\boldsymbol{I}_{[T_k, T_{k+1})}$ denotes the indicator function over the interval $[T_k, T_{k+1})$.

For each $k = 0, \ldots, N$, let B^{H_k} be an n-dimensional fBM with Hurst parameter H_k. Let X be the process defined by

$$X_t = \sum_{j=0}^{k-1} (B_{T_{j+1}}^{H_j} - B_{T_j}^{H_j}) + (B_t^{H_k} - B_{T_k}^{H_k}) \quad \text{whenever} \quad t \in [T_k, T_{k+1}). \tag{7.153}$$

Then X is a continuous process representing an fBM which involves finitely many changes of mode of Hurst parameter (described in (7.152)).

The transition probabilities of the process X are constructed as follows. For each $k = 0, \ldots, N$, let $\theta_k(t) = H_k t^{2H_k - 1}$ for $t \in [T_k, T_{k+1})$. Let $\{p_k(t,x)\}_{k=0}^N$ be a sequence of the unique solutions to the following initial value problems, each defined on $[T_k, T_{k+1}) \times \mathbb{R}^n$:

$$\partial_t p_0(t, x) = \theta_0(t) \Delta p_0(t, x), \ t \in (0, T_1), \ x \in \mathbb{R}^n, \tag{7.154}$$

$$p_0(0, x) = \delta_0(x), \ x \in \mathbb{R}^n, \tag{7.155}$$

where $\delta_0(x)$ is the Dirac delta function with mass on 0, and for $k = 1, \ldots, N$,

$$\partial_t p_k(t, x) = \theta_k(t) \Delta p_k(t, x), \ t \in (T_k, T_{k+1}), \ x \in \mathbb{R}^n, \tag{7.156}$$

$$p_k(T_k, x) = p_{k-1}(T_k-, x), \ x \in \mathbb{R}^n. \tag{7.157}$$

Define functions $\theta(t)$ and $p(t, x)$ respectively by $\theta(t) = \theta_k(t)$ and $p(t, x) = p_k(t, x)$ whenever $t \in [T_k, T_{k+1})$. Then the transition probabilities of X are given by $p(t, x)$ and satisfy

$$\partial_t p(t, x) = \theta(t) \Delta p(t, x), \ t \in \bigcup_{k=0}^N (T_k, T_{k+1}), \ x \in \mathbb{R}^n, \tag{7.158}$$

$$p(0, x) = \delta_0(x), \ x \in \mathbb{R}^n. \tag{7.159}$$

$$p(T_k, x) = p(T_k-, x), \ x \in \mathbb{R}^n, \ k = 1, \ldots, N. \tag{7.160}$$

Discussion of existence and uniqueness of the solution to this type of initial value problem is found in [Umarov and Steinberg (2009)]. The initial value problem associated with the time-changed process $(X_{E_t^\beta})$ is given by

$$D_{*,t}^\beta q(t, x) = \Theta_t^\beta \Delta q(t, x), \ t \in (0, \infty), \ x \in \mathbb{R}^n, \tag{7.161}$$

$$q(0, x) = \delta_0(x), \ x \in \mathbb{R}^n, \tag{7.162}$$

$$q(T_k, x) = q(T_k-, x), \ x \in \mathbb{R}^n, \ k = 1, \ldots, N, \tag{7.163}$$

where the operator Θ_t^β is defined by $\Theta_t^\beta = \sum_{k=0}^N H_k G_{2H_k-1,t}^\beta I_{[T_k, T_{k+1})}(t)$.

Remark 7.12. Combining ideas in Applications 3 and 4, it is possible to construct an fBM having variable Hurst parameter $H(t) \in (1/2, 1)$ with finitely many changes of mode and to establish the associated FPK equations.

Application 5 (Ornstein–Uhlenbeck process). Consider the one-dimensional Ornstein–Uhlenbeck process Y given by

$$Y_t = y_0 e^{-\alpha t} + \sigma \int_0^t e^{-\alpha(t-s)} dB_s, \ t \geq 0, \tag{7.164}$$

where $\alpha \geq 0$, $\sigma > 0$, $y_0 \in \mathbb{R}$ are constants and B is a standard Brownian motion. If $\alpha = 0$, then $Y_t = y_0 + \sigma B_t$, a Brownian motion multiplied by σ starting at y_0. Suppose $\alpha > 0$. The process Y defined by (7.164) is the unique strong solution to the inhomogeneous linear SDE

$$dY_t = -\alpha Y_t dt + \sigma dB_t \ \text{ with } \ Y_0 = y_0, \tag{7.165}$$

which is associated with the SDE

$$d\bar{Y}_t = -\alpha \bar{Y}_t dE_t^\beta + \sigma dB_{E_t^\beta} \ \text{ with } \ \bar{Y}_0 = y_0, \tag{7.166}$$

via the dual relationships $\bar{Y}_t = Y_{E_t^\beta}$ and $Y_t = \bar{Y}_{U_t^\beta}$ (see Theorem 6.2).

Consider the zero-mean process X defined by

$$X_t = Y_t - y_0 e^{-\alpha t} = \sigma \int_0^t e^{-\alpha(t-s)} dB_s = \sigma e^{-\alpha t} \int_0^t e^{\alpha s} dB_s. \qquad (7.167)$$

Note that the Itô integral $(\int_0^t e^{\alpha s} dB_s)_{t \geq 0}$ is a Gaussian process since the integrand $e^{\alpha s}$ is deterministic. This, together with the fact that $\sigma e^{-\alpha t}$ is a deterministic function, implies that X is also a Gaussian process. Direct calculation yields

$$R_X(t) = \frac{\sigma^2}{2\alpha}(1 - e^{-2\alpha t}) \quad \text{and} \quad \widetilde{R_X}(s) = \frac{\sigma^2}{s + 2\alpha}.$$

Therefore, due to Proposition 7.2 and Theorem 7.12, the initial value problems associated with X and $(X_{E_t^\beta})$, where E^β is independent of X, are respectively given by

$$\partial_t p(t, x) = \frac{\sigma^2}{2} e^{-2\alpha t} \partial_x^2 p(t, x), \quad p(0, x) = \delta_0(x); \qquad (7.168)$$

$$D_{*,t}^\beta q(t, x) = \frac{\sigma^2 \beta}{2} J_t^{1-\beta} L_{s \to t}^{-1} \left[\frac{1}{2\pi i} \int_C \frac{\partial_x^2 \tilde{q}(z, x)}{s^\beta - z^\beta + 2\alpha} \, dz \right](t), \quad q(0, x) = \delta_0(x). \qquad (7.169)$$

The unique representation of the solution to the initial value problem (7.168) is obtained via the usual technique using the Fourier transform. Moreover, expression (7.133) guarantees uniqueness of the solution to (7.169) as well.

Notice that the two processes X and $(X_{E_t^\beta})$ are unique strong solutions to SDEs (7.165) and (7.166) with $y_0 = 0$, respectively. Therefore, it is also possible to apply Theorem 7.6 to obtain the following forms of initial value problems which are understood in the sense of generalized functions:

$$\partial_t p(t, x) = \alpha \partial_x \{x p(t, x)\} + \frac{\sigma^2}{2} \partial_x^2 p(t, x), \quad p(0, x) = \delta_0(x); \qquad (7.170)$$

$$D_{*,t}^\beta q(t, x) = \alpha \partial_x \{x q(t, x)\} + \frac{\sigma^2}{2} \partial_x^2 q(t, x), \quad q(0, x) = \delta_0(x). \qquad (7.171)$$

Actually these FPK equations hold in the strong sense as well. For uniqueness of solutions to (7.170) and (7.171), see e.g. [Friedman (1964)] and Corollary 7.5.

The above discussion yields the following two sets of equivalent initial value problems: (7.168) and (7.170), and (7.169) and (7.171). At first glance, PDE (7.168) might seem simpler or computationally more tractable than PDE (7.170); however, PDE (7.169) which is associated with the time-changed process has a more complicated form than PDE (7.171). A significant difference between PDEs (7.168) and (7.170) is the fact that the right-hand side of (7.170) can be expressed as $A^* p(t, x)$ with the *spatial* operator $A^* = \alpha \partial_x x + \frac{\sigma^2}{2} \partial_x^2$ whereas the right-hand side of (7.168) involves both the spatial operator ∂_x^2 and the *time-dependent* multiplication operator by $e^{-2\alpha t}$. This observation suggests: 1) establishing FPK equations for time-changed processes via several different forms of FPK equations for the corresponding untime-changed processes, and 2) choosing appropriate forms for handling specific problems.

Remark 7.13. Since Itô stochastic integrals of deterministic integrands are Gaussian processes, if the variance function of such a stochastic integral satisfies the technical conditions specified in Theorem 7.12, then the FPK equation for the time-changed stochastic integral is explicitly given by (7.130), or equivalently, (7.131).

7.7 Fractional FPK equations associated with stochastic processes which are time changes of solutions of SDEs in bounded domains

This section derives fractional FPK equations associated with stochastic processes in bounded domains. As in Section 4.6, let $\Omega \subset \mathbb{R}^d$ be a bounded domain with a smooth boundary $\partial\Omega \subset \mathbb{R}^{d-1}$ and consider the following initial-boundary value problem considered in (4.32)–(4.34):

$$\frac{\partial u(t,x)}{\partial t} = A(x,D)u(t,x), \quad t > 0, \ x \in \Omega, \tag{7.172}$$

$$\mathcal{W}(x',D)u(t,x') = 0, \quad t > 0, \ x' \in \partial\Omega, \tag{7.173}$$

$$u(0,x) = u_0(x), \quad x \in \Omega, \tag{7.174}$$

where $A(x,D)$ is a second order Waldenfels operator given in (4.35) and $\mathcal{W}(x',D)$ is a boundary pseudo-differential operator given in (4.36). As in Section 7.2, let $U(\mu,t) \in \mathcal{S}$ denote a strictly increasing subordinator satisfying

$$\mathbb{E}[e^{-sU(\mu,t)}] = e^{-t\Phi_\mu(s)} \quad \text{with} \quad \Phi_\mu(s) = \int_0^1 s^\beta d\mu(\beta) \tag{7.175}$$

for some finite positive measure μ on $(0,1)$.

Theorem 7.14. *Let X_t be a stochastic process associated with the FPK equation (7.172)–(7.174) and $E(\mu,t)$ be the inverse to a subordinator $U(\mu,t) \in \mathcal{S}$ with a mixing measure μ which is independent of X_t. Then the FPK equation associated with the time-changed stochastic process $X_{E(\mu,t)}$ has the form*

$$D_\mu v(t,x) = A(x,D)v(t,x), \quad t > 0, \ x \in \Omega, \tag{7.176}$$

$$\mathcal{W}(x',D)v(t,x') = 0, \quad t > 0, \ x' \in \partial\Omega, \tag{7.177}$$

$$v(0,x) = u_0(x), \quad x \in \Omega, \tag{7.178}$$

where $u_0 \in C^2_\mathcal{W}(\Omega) \equiv \{\varphi \in C^2(\Omega) : \mathcal{W}(x',D)\varphi(x') = 0 \text{ if } x' \in \partial\Omega\}$.

Proof. Let $\{T_t\}$ be the semigroup with the infinitesimal generator $A = A(x,D) : C(\overline{\Omega}) \to C(\overline{\Omega})$, with the domain

$$\text{Dom}(A) = \{\phi \in C^2(\Omega) : \mathcal{W}(x',D)\phi(x') = 0, \ x' \in \partial\Omega\}.$$

Then the unique solution of the problem (7.172)–(7.174) has the form

$$u(t, x) = T_t u_0(x) = \mathbb{E}[u_0(X_t)|X_0 = x],$$

indicating connection of the solution $u(t, x)$ of the FPK equation in (7.172)–(7.174) with the stochastic process X_t. Now, consider the function $v(t, x)$ obtained from the latter replacing X_t by $X_{E(\mu,t)}$. Since X_t and $E(\mu, t)$ are independent, $v(t, x)$ has the following form:

$$
\begin{aligned}
v(t, x) &= \mathbb{E}[u_0(X_{E(\mu,t)})|X_0 = x] \\
&= \int_0^\infty \mathbb{E}[u_0(X_{E(\mu,t)})|E(\mu, t) = \tau, X_0 = x] f_{E(\mu,t)}(\tau) d\tau \\
&= \int_0^\infty u(\tau, x) f_{E(\mu,t)}(\tau) d\tau \\
&= \int_0^\infty f_{E(\mu,t)}(\tau) T_\tau u_0(x) d\tau.
\end{aligned}
\tag{7.179}
$$

We will show that $v(t, x)$ defined above satisfies the initial-boundary value problem for the fractional FPK equation in (7.176)–(7.178). First, we show that $v(t, x)$ satisfies equation (7.176). Indeed, one can readily see that

$$v(t, x) = -\int_0^\infty \frac{\partial}{\partial \tau} \{J f_{U(\mu,\tau)}(t)\} (T_\tau u_0(x)) d\tau.$$

Now it follows from the definition of $U(\mu, t)$ along with part (d) of Lemma 7.3 that the Laplace transform of $v(t, x)$ satisfies

$$
\begin{aligned}
\tilde{v}(s, x) &= \frac{\int_0^1 s^\beta d\mu(\beta)}{s} \int_0^\infty e^{-\tau \int_0^1 s^\beta d\mu(\beta)} (T_\tau u_0(x)) d\tau \\
&= \frac{\Phi_\mu(s)}{s} \tilde{u}(\Phi_\mu(s), x), \quad s > \bar{\omega},
\end{aligned}
\tag{7.180}
$$

where $\tilde{u}(s, x)$ is the Laplace transform of $u(t, x)$, the function $\Phi_\mu(s)$ is defined in (7.175), and $\bar{\omega} > 0$ is a number such that $s > \bar{\omega}$ if $\Phi_\mu(s) > \omega$ ($\bar{\omega}$ is uniquely defined since, due to (7.175), the function $\Phi_\mu(s)$ is a strictly increasing function). The function $\tilde{u}(s, x)$ satisfies the equation

$$s\tilde{u}(s, x) - A(x, D)\tilde{u}(s, x) = u_0(x), \quad x \in \Omega,
\tag{7.181}$$

Indeed, applying the Laplace transform to both sides of equation (7.172) and taking into account the initial condition (7.174) yields equation (7.181). It follows from equations (7.180) and (7.181) that the composite function

$$\tilde{u}(\Phi_\mu(s), x) = \frac{s\tilde{v}(s, x)}{\Phi_\mu(s)}$$

satisfies the equation

$$\left(\Phi_\mu(s) - A(x, D)\right) \frac{s\tilde{v}(s, x)}{\Phi_\mu(s)} = u_0(x), \quad s > \bar{\omega}, \; x \in \Omega,$$

or equivalently, the equation

$$\Big(\Phi_\mu(s) - A(x, D)\Big)\tilde{v}(s, x) = u_0(x)\,\frac{\Phi_\mu(s)}{s}, \quad s > \bar{\omega},\ x \in \Omega.$$

Rewrite the latter in the form

$$\Phi_\mu(s)\tilde{v}(s, x) - \frac{\Phi_\mu(s)}{s}v(0+, x) = A(x, D)\tilde{v}(s, x), \quad s > \bar{\omega},\ x \in \Omega. \tag{7.182}$$

Notice that the left hand side of the latter equation is the Laplace transform of the expression $D_\mu v(t, x)$ due to formula (3.31). Therefore, equation (7.182) is equivalent to equation (7.176).

Further, using (4.33), we have

$$\mathcal{W}(x', D)v(t, x') = \mathcal{W}(x', D) \int_0^\infty u(\tau, x')f_{E(\mu,t)}(\tau)d\tau$$

$$= \int_0^\infty \mathcal{W}(x', D)u(\tau, x')f_{E(\mu,t)}(\tau)d\tau = 0$$

since $\mathcal{W}(x', D)u(\tau, x') = 0$ for all $\tau > 0$ due to the boundary condition (4.33).

Finally, since X_t is right-continuous and Ω is a bounded domain, the dominated convergence theorem yields

$$v(0, x) = \lim_{t \to 0+} \mathbb{E}[u_0(X_{E(\mu,t)})|X_0 = x] = \mathbb{E}[u_0(X_0)|X_0 = x] = u_0(x).$$

Hence, $v(t, x)$ defined in (7.179) satisfies the initial-boundary value problem in (7.176)–(7.178) for the fractional order FPK equation. □

In the particular case of E_t^β being the inverse of a single stable subordinator U_t^β with stability index $\beta \in (0, 1)$, this theorem implies the following result:

Corollary 7.15. *Let X_t be a stochastic process associated with the FPK equation (7.172)–(7.174) and E_t^β be the inverse to a stable subordinator with index $0 < \beta < 1$ which is independent of X_t. Then the FPK equation associated with the time-changed stochastic process $X_{E_t^\beta}$ has the form*

$$D_*^\beta v(t, x) = A(x, D)v(t, x), \quad t > 0,\ x \in \Omega, \tag{7.183}$$

$$\mathcal{W}(x', D)v(t, x') = 0, \quad t > 0,\ x' \in \partial\Omega, \tag{7.184}$$

$$v(0, x) = u_0(x), \quad x \in \Omega. \tag{7.185}$$

An important question is the existence of a unique solution of the initial-boundary value problem in equations (7.183)–(7.185).

Theorem 7.16. *Assume that the Waldenfels operator $A(x, D)$ in (7.176) and Wentcel's boundary operator $W(t, x')$ in (7.177) satisfy conditions $(i) - (iv)$, $(a) - (d)$, and $(C1) - (C3)$ in Section 4.6. Then the initial-boundary value problem (7.176)–(7.178) for the fractional distributed order FPK equation has a unique solution $v(t, x)$ in the space $C([0, \infty) \times \bar{\Omega}) \cap C^1(t > 0; C^2_{\mathcal{W}}(\Omega))$, where $C^1(t > 0; C^2_{\mathcal{W}}(\Omega))$ is the space of vector-functions differentiable in t and belonging to $C^2_{\mathcal{W}}(\Omega)$ for each fixed $t > 0$.*

Proof. The proof of this theorem follows from the representation of the solution of initial boundary value problem (7.176)–(7.178) in the form (see equation (7.179))

$$v(t,x) = \int_0^\infty f_{E(\mu,t)}(\tau)u(\tau,x)d\tau, \qquad (7.186)$$

where $u(t,x)$ is the unique solution of initial-boundary value problem (7.172)–(7.174). We proved in Theorem 4.6 that $u(t,x)$ belongs to the space $C([0,\infty) \times \bar{\Omega}) \cap C^1(t > 0; C^2_{\mathcal{W}}(\Omega))$. It follows from this fact and representation (7.186) that $v(t,x)$ has all derivatives if $t > 0$ and that the estimate

$$|v(t,x)| \leq \int_0^\infty f_{E(\mu,t)}(\tau)|T_\tau u_0(x)|d\tau \leq \sup_{t \geq 0}\|T_t u_0(x)\|_{C(\bar{\Omega})}$$

$$= \sup_{t \geq 0}\|u(t,x)\|_{C(\bar{\Omega})}, \quad t \geq 0, \ x \in \bar{\Omega},$$

holds. Thus, the function $v(t,x)$ inherits all the properties of $u(t,x)$, including being in the space $C([0,\infty) \times \bar{\Omega}) \cap C^1(t > 0; C^2_{\mathcal{W}}(\Omega))$.

Now, assume there are two distinct solutions $v_1(t,x)$ and $v_2(t,x)$ of the problem (7.172)–(7.174) belonging to $C([0,\infty) \times \bar{\Omega}) \cap C^1(t > 0; C^2_{\mathcal{W}}(\Omega))$. Let $w(t,x) = v_1(t,x) - v_2(t,x)$. Then $w(t,x)$ also belongs to the space $C([0,\infty) \times \bar{\Omega}) \cap C^1(t > 0; C^2_{\mathcal{W}}(\Omega))$. The function $w(t,x)$ satisfies initial-boundary value problem (7.176)–(7.178) with $u_0(x) \equiv 0$ in initial condition (7.178). Then, as was established in Theorem 4.6, $u(t,x) \equiv 0$. Due to representation (7.186), this in turn implies $w(t,x) \equiv 0$. Thus $v_1(t,x) \equiv v_2(t,x)$, thereby establishing uniqueness of the solution. $\qquad \square$

Now consider the following initial-boundary value problem (7.172)–(7.174) for t-dependent generalizations of the FPK equation

$$\frac{\partial u(t,x)}{\partial t} = B(x,D)u(t,x) + \frac{(\gamma+1)t^\gamma}{2}A(x,D)u(t,x), \quad t > 0, \ x \in \Omega, \qquad (7.187)$$

$$W(x',D)u(t,x') = 0, \quad t > 0, \ x' \in \partial\Omega, \qquad (7.188)$$

$$u(0,x) = \varphi(x), \quad x \in \Omega, \qquad (7.189)$$

where $B(x,D)$ is a pseudo-differential operator whose order is strictly less than the order of the operator $A(x,D)$. We assume that $A(x,D)$ is an elliptic Waldenfels operator defined in (4.35) and $W(x',D)$ is Wentcel's boundary pseudo-differential operator defined in (4.36). The parameter γ is in the interval $(-1,1)$. Equation (7.187) is a parabolic equation: if $0 < \gamma < 1$, then it is degenerate; if $-1 < \gamma < 0$, then it is singular. Obviously, the initial-boundary value problem (7.187)–(7.189) recovers problem (7.172)–(7.174) if $B(x,D) = 0$ and $\gamma = 0$.

Consider the following motivating example: let $B(x,D) = 0$, $\gamma = 2H - 1$ and $A(x,D) = \Delta$, the Laplace operator. Then equation (7.187) reduces to

$$\frac{\partial u(t,x)}{\partial t} = Ht^{2H-1}\Delta u(t,x), \quad t > 0, \ x \in \Omega,$$

which represents the FPK equation associated with fractional Brownian motion (fBM) with Hurst parameter $H \in (0,1)$. Therefore, the initial-boundary value problem (7.187)–(7.189) can be considered as an FPK equation of a stochastic process in a bounded region Ω driven not only by Lévy processes but also by fBM. Below we derive the fractional FPK equation associated with such a stochastic process with a time-changed driving process.

Theorem 7.17. *Let the density function $u(t, x)$ of a stochastic process X_t satisfy the FPK equation (7.187)–(7.189). Let $E(\mu, t)$ be the inverse to a subordinator $U(\mu, t) \in S$ with a mixing measure μ which is independent of X_t. Then the density function $v(t, x)$ of the time-changed process $X_{E(\mu,t)}$ satisfies the FPK equation*

$$D_\mu v(t, x) = B(x, D)v(t, x) + \frac{\gamma + 1}{2} G_{\gamma,t}^\mu A(x, D)v(t, x), \quad t > 0, \ x \in \Omega, \quad (7.190)$$

$$\mathcal{W}(x', D)v(t, x') = 0, \quad t > 0, \ x' \in \partial\Omega, \quad (7.191)$$

$$v(0, x) = u_0(x), \quad x \in \Omega, \quad (7.192)$$

where the operator $G_{\gamma,t}^\mu$ is defined as

$$G_{\gamma,t}^\mu v(t, x) = \mathcal{K}_\mu(t) * L_{s \to t}^{-1} \left[\frac{\Gamma(\gamma + 1)}{2\pi i} \int_{C-i\infty}^{C+i\infty} \frac{m_\mu(z)\tilde{v}(z, x)}{(\Phi_\mu(s) - \Phi_\mu(z))^{\gamma+1}} dz \right](t), \quad (7.193)$$

where $*$ denotes the convolution operation, the symbol $L_{s \to t}^{-1}$ means the inverse Laplace transform, $0 < C < s$, the functions $\Phi_\mu(z)$ and $m_\mu(z)$ are defined by

$$\Phi_\mu(z) = \int_0^1 e^{\beta Ln(z)} d\mu(\beta), \quad m_\mu(z) = \frac{\int_0^1 \beta z^\beta d\mu(\beta)}{\Phi_\mu(z)},$$

the function $\mathcal{K}_\mu(t)$ is defined in (7.78), and $u_0 \in C_{\mathcal{W}}^2(\Omega) \equiv \{\phi \in C^2(\Omega) : \mathcal{W}(x', D)\phi(x') = 0 \ if \ x' \in \partial\Omega\}$.

Proof. By independence between X_t and $E(\mu, t)$,

$$v(t, x) = \int_0^\infty f_{E(\mu,t)}(\tau)u(\tau, x)d\tau, \quad t \geqslant 0, \ x \in \Omega. \quad (7.194)$$

By assumption, the function $u(t, x)$ solves problem (7.187)–(7.189). We will show that $v(t, x)$ satisfies problem (7.190)–(7.192). To show that $v(t, x)$ satisfies equation (7.190), we compute

$$D_\mu v(t, x) = \int_0^\infty D_\mu f_{E(\mu,t)}(\tau)u(\tau, x)d\tau.$$

Here the change of the order of D_μ and the integral is valid due to the estimate obtained in [Hahn and Umarov (2011)] for the density function $f_{E(\mu,t)}(\tau)$ of a mixture of stable subordinators having mixing measure μ. It follows from Lemma 7.4 that

$$D_\mu v(t, x) = -\int_0^\infty \frac{\partial f_{E(\mu,t)}(\tau)}{\partial \tau} u(\tau)d\tau - \mathcal{K}_\mu(t) \int_0^\infty \delta_0(\tau)u(\tau)d\tau. \quad (7.195)$$

Integrating by parts in the first integral yields

$$-\int_0^\infty \frac{\partial f_{E(\mu,t)}(\tau)}{\partial \tau} u(\tau) d\tau = \int_0^\infty f_{E(\mu,t)}(\tau) \frac{\partial u(\tau)}{\partial \tau} d\tau - \lim_{\tau \to \infty} f_{E(\mu,t)}(\tau) u(\tau)$$

$$+ \lim_{\tau \to 0} f_{E(\mu,t)}(\tau) u(\tau). \tag{7.196}$$

The first limit on the right hand side is zero due to part (c) of Lemma 7.3. Due to part (b) of Lemma 7.3 the second limit on the right hand side of (7.196) has the same value as the second integral on the right side of (7.195), but with the opposite sign. Hence, it follows that

$$D_\mu v(t,x) = \int_0^\infty f_{E(\mu,t)}(\tau) \frac{\partial}{\partial \tau} u(\tau,x) d\tau.$$

Now using equation (7.187), we have

$$D_\mu v(t,x) = \int_0^\infty f_{E(\mu,t)}(\tau) \left[B(x,D) u(\tau,x) + \frac{(\gamma+1)\tau^\gamma}{2} A(x,D) u(\tau,x) \right] d\tau$$

$$= B(x,D) \int_0^\infty f_{E(\mu,t)}(\tau) u(\tau,x) d\tau + \frac{\gamma+1}{2} A(x,D) \int_0^\infty f_{E(\mu,t)}(\tau) \tau^\gamma u(\tau,x) d\tau$$

$$= Bv(t,x) + \frac{\gamma+1}{2} A(x,D) \int_0^\infty f_{E(\mu,t)}(\tau) \tau^\gamma u(\tau,x) d\tau$$

$$= Bv(t,x) + \frac{\gamma+1}{2} A G^\mu_{\gamma,t} v(t,x),$$

where

$$G^\mu_{\gamma,t} v(t,x) = \int_0^\infty f_{E(\mu,t)}(\tau) \tau^\gamma u(\tau,x) d\tau. \tag{7.197}$$

The fact that the operator $G^\mu_{\gamma,t}$ has the representation (7.193) is verifed in the proof of Theorem 7.9.

It is easy to verify that $\mathcal{W}(x',D) v(t,x') = 0$ if $x' \in \partial\Omega$. Indeed, using $\mathcal{W}(x',D) u(\tau,x') = 0$ for all $\tau \geq 0$, we have

$$\mathcal{W}(x',D) v(t,x') = \int_0^\infty \mathcal{W}(x',D) u(\tau,x') f_{E(\mu,t)}(\tau) d\tau = 0, \quad x' \in \partial\Omega,$$

for any fixed $t \geq 0$.

Finally, making use of part (a) of Lemma 7.3 and the dominated convergence theorem,

$$\lim_{t \to 0+} v(t,x) = \lim_{t \to 0+} \int_0^\infty f_{E(\mu,t)}(\tau) u(\tau,x) d\tau$$

$$= \int_0^\infty \lim_{t \to 0+} f_{E(\mu,t)}(\tau) \, u(\tau,x) d\tau$$

$$= \int_0^\infty \delta_0(\tau) \, u(\tau,x) d\tau$$

$$= u(0,x) = u_0(x).$$

Hence, $v(t,x)$ defined in (7.194) satisfies the initial-boundary value problem in (7.190)–(7.192) for the time-dependent fractional order FPK equation. □

Remark 7.14. The properties of the operator $G_{\gamma,t}^{\mu}$ are studied in Section 7.4, including the fact that the family $\{G_{\gamma,t}^{\mu}, -1 < \gamma < 1\}$ possesses the semigroup property. Namely, for any $\gamma, \delta \in (-1, 1)$ with $\gamma + \delta \in (-1, 1)$, the identity $G_{\gamma,t}^{\mu} \circ G_{\delta,t}^{\mu} = G_{\gamma+\delta,t}^{\mu} = G_{\delta,t}^{\mu} \circ G_{\gamma,t}^{\mu}$ holds, where "$\circ$" denotes the composition of two operators.

As in Corollary 7.15, in the particular case of E_t^{β} being the inverse of a single stable subordinator U_t^{β} with index $\beta \in (0, 1)$, Theorem 7.17 implies the following result:

Corollary 7.18. *Let the density function $u(t, x)$ of a stochastic process X_t satisfy the FPK equation (7.187)–(7.189). Let E_t^{β} be the inverse to a stable subordinator with index $0 < \beta < 1$ which is independent of X_t. Then the density function $v(t, x)$ of the time-changed process $X_{E(\mu,t)}$ satisfies the FPK equation*

$$D_*^{\beta} v(t, x) = B(x, D)v(t, x) + \frac{\gamma + 1}{2} G_{\gamma,t}^{\mu} A(x, D)v(t, x), \quad t > 0, \ x \in \Omega,$$

$$\mathcal{W}(x', D)v(t, x') = 0, \quad t > 0, \ x' \in \partial\Omega,$$

$$v(0, x) = u_0(x), \quad x \in \Omega.$$

Theorem 7.19. *Let the operators $A(x, D)$, $B(x, D)$, and $\mathcal{W}(x', D)$ in problem (7.190)–(7.192) satisfy the following conditions:*

- *(A) The pseudo-differential operator $A(x, D)$ is an elliptic Waldenfels operator defined in (4.35) and satisfying the conditions (i)–(iv) and (C1) in Section 4.6;*
- *(B) The operator $B(x, D)$ is a pseudo-differential operator whose order is strictly less than the order of $A(x, D)$;*
- *(W) The pseudo-differential operator $\mathcal{W}(x', D)$ is Wentcel's boundary operator defined in (4.36) and satisfying the conditions (a)–(d), (C2) and (C3) in Section 4.6.*

Then for an arbitrary $u_0 \in C_{\mathcal{W}}^2(\Omega)$, the initial-boundary value problem (7.190)–(7.192) has a unique solution $v(t, x)$ in the space $C([0, \infty) \times \bar{\Omega}) \cap C^1(t > 0; C_{\mathcal{W}}^2(\Omega))$.

Proof. The argument given in the proof of Theorem 7.16 does not work in the case of the problem (7.190)–(7.192) since the solution $u(t, x)$ of the problem (7.187)–(7.189) does not have a representation through the Feller semigroup. To prove this theorem, we will use properties of pseudo-differential operators.

Consider the operator

$$\mathcal{A}_{\mathcal{W}}(t) = \frac{t^{\gamma}}{2} A(x, D) + B(x, D)$$

with the domain $\text{Dom}(\mathcal{A}_{\mathcal{W}}(t)) = \{\phi \in C^2(\overline{\Omega}) : \mathcal{W}(x', D)\phi(x') = 0, \ x' \in \partial\Omega\}$. This operator is a pseudo-differential operator with the symbol

$$\sigma(t, x, \xi) = \frac{t^{\gamma}}{2} \sigma_A(x, \xi) + \sigma_B(x, \xi), \quad t \geq 0, \ x \in \Omega, \ \xi \in \mathbb{R}^d, \tag{7.198}$$

where $\sigma_A(x, \xi)$ and $\sigma_B(x, \xi)$ are symbols of the operators $A(x, D)$ and $B(x, D)$, respectively. Due to conditions (A) and (B) of the theorem, for each fixed $t > 0$, the symbol $\sigma(t, x, \xi)$ satisfies the following ellipticity estimate

$$- \sigma(t, x, \xi) \geqslant \kappa_t |\xi|^\delta, \quad |\xi| \geqslant C, \tag{7.199}$$

with some constants $\kappa_t > 0$, $C > 0$, and $\delta > 0$.

One can see that the solution of the problem (7.187)–(7.189) has a formal representation

$$u(t, x) = S(t, x, D) u_0(x), \quad t \geqslant 0, \ x \in \Omega, \tag{7.200}$$

where the solution pseudo-differential operator $S(t, x, D)$ has the symbol

$$s(t, x, \xi) = e^{t\sigma(t, x, \xi)}, \quad t \geqslant 0, \ \xi \in \mathbb{R}^d. \tag{7.201}$$

The fact that $u(t, x)$ satisfies equation (7.187) can be verified by direct calculation. To show this, let us extend $u_0(x)$ to be 0 for all $x \in \mathbb{R}^d \backslash \Omega$, and denote the extended function again by $u_0(x)$. Let $\tilde{u}_0(\xi)$ be the Fourier transform of $u_0(x)$. Then

$$\begin{aligned}
\frac{\partial u(t, x)}{\partial t} &= \frac{\partial S(t, x, D) u_0(x)}{\partial t} \\
&= \frac{1}{(2\pi)^d} \frac{\partial}{\partial t} \int_{\mathbb{R}^d} e^{t\sigma(t, x, \xi)} e^{-i(x, \xi)} \tilde{u}_0(\xi) d\xi \\
&= \frac{1}{(2\pi)^d} \int_{\mathbb{R}^d} \left[\sigma(t, x, \xi) + t \frac{\partial \sigma(t, x, \xi)}{\partial t} \right] e^{t\sigma(t, x, \xi)} e^{-i(x, \xi)} \tilde{u}_0(\xi) d\xi \\
&= \frac{1}{(2\pi)^d} \int_{\mathbb{R}^d} \left[\frac{(\gamma + 1) t^\gamma}{2} \sigma_A(x, \xi) + \sigma_B(x, \xi) \right] e^{-i(x, \xi)} e^{t\sigma(t, x, \xi)} \tilde{u}_0(\xi) d\xi \\
&= \left[\frac{(\gamma + 1) t^\gamma}{2} A(x, D) + B(x, D) \right] w(t, x), \quad t > 0, \ x \in \Omega,
\end{aligned}$$

where $w(t, x)$ has the Fourier transform

$$\tilde{w}(t, \xi) = e^{t\sigma(t, x, \xi)} \tilde{u}_0(\xi), \quad t > 0, \ \xi \in \mathbb{R}^d.$$

Changing the differentiation and integration operators in the above calculation is legitimate due to estimate (7.199). Now calculating the inverse Fourier transform of $\tilde{w}(t, \xi)$, and using the definition (4.2) of pseudo-differential operators,

$$\begin{aligned}
w(t, x) &= \frac{1}{(2\pi)^d} \int_{\mathbb{R}^d} e^{t\sigma(t, x, \xi)} e^{-i(x, \xi)} \tilde{u}_0(\xi) d\xi \\
&= \frac{1}{(2\pi)^d} \int_{\mathbb{R}^d} s(t, x, \xi) e^{-i(x, \xi)} \tilde{u}_0(\xi) d\xi \\
&= S(t, x, D) u_0(x), \quad t > 0, \ x \in \Omega.
\end{aligned}$$

Thus, $w(t, x) = u(t, x)$, and hence, $u(t, x)$ defined by (7.200) satisfies equation (7.187). Moreover, since $W(x', D) u_0(x') = 0$ for $x' \in \partial \Omega$, it follows from (7.200) that $W(x', D) u(t, x') = 0$ for all $t > 0$ and $x' \in \partial \Omega$. It also follows from (7.201) that $S(0, x, D) = I$, the identity operator, implying $u(0, x) = u_0(x)$. Finally, estimate

(7.199) and representation (7.200) of the solution imply that the solution belongs to the space $C([0, \infty) \times \bar{\Omega}) \cap C^1(t > 0; C^2_{\mathcal{W}}(\Omega))$.

The existence of a solution to the initial-boundary value problem (7.190)–(7.192) immediately follows from representation (7.194), namely,

$$v(t, x) = \int_0^\infty f_{E(\mu,t)}(\tau) u(\tau, x) d\tau, \tag{7.202}$$

where $u(t, x)$ is the solution of the problem (7.187)–(7.189). The argument here is similar to the proof of Theorem 7.16. Also, as in Theorem 7.16, the function $v(t, x)$ inherits all the properties of $u(t, x)$, including being in $C([0, \infty) \times \bar{\Omega}) \cap C^1(t > 0; C^2_{\mathcal{W}}(\Omega))$.

Finally, to show the uniqueness of the solution, we assume that there exist two distinct solutions $v_1(t, x)$ and $v_2(t, x)$ of problem (7.190)–(7.192) in the space $C([0, \infty) \times \bar{\Omega}) \cap C^1(t > 0; C^2_{\mathcal{W}}(\Omega))$. Let $w(t, x) = v_1(t, x) - v_2(t, x)$. Obviously, $w(t, x)$ belongs to the space $C([0, \infty) \times \bar{\Omega}) \cap C^1(t > 0; C^2_{\mathcal{W}}(\Omega))$ and solves the problem (7.190)–(7.192) with $u_0(x) \equiv 0$ in the initial condition (7.192). Further, equation (7.200) shows that $u(t, x) \equiv 0$ in this case. This, together with representation (7.202), implies $w(t, x) \equiv 0$. Hence, $v_1(t, x) \equiv v_2(t, x)$, implying the uniqueness of the solution. $\square$

Remark 7.15. The technique used to prove Theorem 7.19 is applicable to FPK equations in the whole space $\mathbb{R}^d$ obtained in the previous sections.

Final Note

To those interested in the theory or application of stochastic processes, stochastic differential equations, or fractional order Fokker–Planck–Kolmogorov equations:

Our hope is that this book has stimulated you to think about the information and insights that might be gained from considering the interconnections in the paradigm below, rather than its vertices or center in isolation. This book should provide the beginning for many investigations and applications, rather than the end.

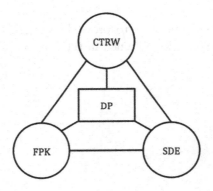

Bibliography

Abdel-Rehim, E.A.(2013) Explicit approximation solutions and proof of convergence of the space-time fractional advection dispersion equations. *Appl. Math.*, **4**, 1427–1440.

Abel, N.H. (1826) Solution of a mechanical problem. (Translated from the German) In: D. E. Smith (ed), *A Source Book in Mathematics.* Dover Publications, New York, 656–662, 1959.

Abramowitz, M., Stegun, I. (1972) *Handbook of Mathematical Functions with Formulas, Graphs, and Mathematical Tables.* Dover Publications, New York.

Adler, R., Feldman, R., Taqqu, M. (1998) *A Practical Guide to Heavy Tails.* Birkhäuser, Boston.

Alòs, E., Mazet, O., Nualart, D. (2001) Stochastic calculus with respect to Gaussian processes. *Ann. Probab.* **29** (2), 766–801.

Anderson, P., Meerschaert, M.M. (1998) Modeling river flows with heavy tails. *Water Resour. Res.* **34**, 2271–2280.

Andries, E., Umarov, S., Steinberg, S. (2006) Monte-Carlo random walk simulations based on distributed order differential equations with applications to cell biology. *Fract. Calc. Appl. Anal,* **9** (4) 351–369.

Anh, V.V., McVinish, R. (2003) Fractional differential equations driven by Lévy noise. *J. Appl. Math. Stoch. Anal.*, **16** (2), 97–119.

Applebaum, D. (2009) *Lévy Processes and Stochastic Calculus*, 2nd edition. Cambridge University Press.

Bachelier, L. (1900) Théorie de la spéculation. *Annales scientifiques de l'École Normale Supérieure, Ser 3* **17**, 21–86. [English translation can be find online at http://www.radio.goldseek.com/bachelier-thesis-theory-of-speculation-en.pdf]

Baeumer, B., Meerschaert, M.M. (2001) Stochastic solutions for fractional Cauchy problems. *Fract. Calc. Appl. Anal.* **4**, 481–500.

Baeumer, B., Meerschaert, M.M. (2010) Tempered stable Lévy motion and transient superdiffusion. *J. Comput. Appl. Math.* **233**, 2438–2448.

Baeumer, B., Meerschaert, M.M., Nane, E. (2009) Brownian subordinators and fractional Cauchy problems. *Trans. Amer. Math. Soc.* **361**, 3915–3930.

Baeumer, B., Kurita, S., Meerschaert, M M. (2005a) Inhomogeneous fractional diffusion equations. *Fract. Calc. Appl. Anal.* **8** (4) 371–386.

Baeumer, B., Meerschaert, M.M. Mortensen, J. (2005b) Space-time fractional derivative operators. *Proc. Amer. Math. Soc.* **133** (8) 2273–2282.

Bagley, R.L., Torvik, P.J. (2000) On the existence of the order domain and the solution of distributed order equations. P 1, 2. *Int. J. Appl. Math.* **2** (7 and 8) 865–882, 965–987.

Barndorf-Nielsen, O.E., Mikosch, T., Resnick, S.I., editors. (2001) *Lévy Processes: Theory and Applications.* Birkhäuser.

Baudoin, F., Coutin, L. (2007) Operators associated with a stochastic differential equation driven by fractional brownian motions, *Stoch. Proc. Appl.* **117** (5) 550–574.

Bazhlekova, E.G. (1998) The abstract Cauchy problem for the fractional evolution equation. *Fract. Calc. Appl. Anal,* **1**, 255–270.

Bazhlekova, E.G. (2000) Subordination principle for fractional evolution equations. *Fract. Calc. Appl. Anal.* **3** (3) 213–230.

Bell, D.R. (1995) *Degenerate Stochastic Differential Equations and Hypoellipticity.* Longman, Harlow.

Bender, C. (2003) An Itô formula for generalized functionals of a fractional Brownian motion with arbitrary Hurst parameter. *Stoch. Proc. Appl.* **104** (1), 81–106.

Benson, D.A., Wheatcraft, S.W., Meerschaert, M.M. (2000) Application of a fractional advection-dispersion equation. *Water Resour. Res.* **36** (6), 1403–1412.

Bertoin, J. (1996) *Lévy Processes.* Cambridge University Press.

Biagini, F., Hu, Y., Øksendal, B., Zhang, T. (2008) *Stochastic Calculus for Fractional Brownian Motion and Applications.* Springer.

Billingsley, P. (1999) *Convergence of Probability Measures.* Wiley Series in Prob. and Stat.

Bingham, N.H., Goldie, C.M., Teugels, J.L. (1987) *Regular Variation.* Cambridge University Press.

Black, F., Scholes M. (1973) The pricing of options and corporate liabilities. *J. Polit. Economy,* **3**, 637–659.

Blumenthal, R.M., Getoor, R.K. (1961) Sample functions of stochastic processes with stationary independent increments. *J. Math. Mech.* **10**, 493–516.

Borland, L. (1998) Microscopic dynamics of the nonlinear Fokker-Planck equations: A phenomenological model. *Phys. Rev. E* **57**, 6634–6642.

Bouchaud, J., Georges, A. (1990) Anomalous diffusion in disordered media: Statistical mechanisms, models and physical applications, *Phys. Rep.,* **195**, 127–293.

Boyer, D., Miramontes, O., Ramos-Fernandez, G., Mateos, G.L., Cocho, G. (2004) Modeling the searching behavior of social monkeys. *Physica A: Statistical Mechanics and its Applications* **342** (1-2) 329–335.

Burr, M.N. (2011) Weak convergence of stochastic integrals driven by continuous-time random walks. `arXiv:1110.0216` [math.PR]

Caputo, M. (1969) *Elasticitá e Dissipazione.* Zanichelli, Bologna.

Caputo, M. (1967) Linear models of dissipation whose Q is almost frequency independent, Part 2. *Geophys. J. R. Astr. Soc.* **13**, 529–539.

Caputo, M. (1995) Mean fractional order derivatives. Differential equations and filters. *Annals Univ. Ferrara - Sez.* VII - SC. Mat., **XLI** 73–84.

Caputo, M. (2001) Distributed order differential equations modeling dielectric induction and diffusion. *Fract. Calc. Appl. Anal.,* **4**, 421–442.

Cartea, Á., Del-Castillo-Negrete, D. (2007) Fractional diffusion models of option prices in markets with jumps. *Physica A,* **374** (2), 749–763.

Chechkin, A.V., Gorenflo, R., Sokolov, I.M., Gonchar, V.Yu. (2003) Distributed order time fractional diffusion equation. *Fract. Calc. Appl. Anal.,* **6**, 259–279.

Chechkin, A.V., Gorenflo, R., Sokolov, I.M. (2005) Fractional diffusion in inhomogeneous media. *J. Physics. A: Math. Gen.*, **38**, L679–L684.

Cheridito, P. (2003) Arbitrage in fractional Brownian motion models. *Finance Stochast.***7**, 533–553.

Cheridito, P. (2001) Mixed fractional Brownian motion. *Bernoulli.* **7**, 913–934.

Ciesielski, M., Leszczynski, J. (2003) Numerical simulations of anomalous diffusion. *Computer Methods in Mechanics,* June 3–6, Gliwice, Poland.

Cont, R., Tankov, P. (2003) *Financial Modelling with Jump Processes.* Chapman and Hall/CRC.

Courrége, P. (1964) Générateur infinitésimal d'un semi-groupe de convolution sur $\mathbb{R}^n$, et formule de Lévy-Khinchine. *Bull. Sci. Math. (2)* **88**, 3–30.

Coutin, L., Decreusefond, L. (1997) Stochastic differential equations driven by fractional Brownian motion. *Tech. Report.*

Decreusefond, L. (2005) Stochastic Integration with respect to Volterra processes. *Annales de l'Institut H. Poincaré*, **41**(2), 123–149.

Decreusefond, L., Uštünel, A.S. (1998) Stochastic Analysis of the Fractional Brownian Motion. *Potential Analysis,* **10** (2), 177–214.

Diethelm K., Ford, N.J. (2001) Numerical solution methods for distributed order differential equations. *Fract. Calc. Appl. Anal.*, **4**, 531–542.

Djrbashian, M.M. (1966) *Integral Transforms and Representations of Functions in the Complex Plane.* Moscow, Nauka (in Russian).

Dolcy, D.S.L., Constantinou, C.C., Quigley, S.F. (2007) A Fokker-Planck equation method predicting Buffer occupancy in a single queue. *Computer Networks,* **51** (8) 2198–2216.

Dubinskii, Yu.A. (1982) The algebra of pseudo-differential operators with analytic symbols and its applications to mathematical physics. *Soviet Math. Survey,* **37**, 107–153.

Dubinskii, Yu.A. (1991) *Analytic Pseudo-differential Operators and Their Applications.* Dordrecht, Kluwer Academic Publishers.

Dudley, R.M. (1973) Sample functions of the Gaussian process. *Ann. Probab.*, **1** (1), 66–103.

Dudley, R.M., Norvaisa, R. (1998) *An Introduction to P-variation and Young Integrals.* MaPhysto. Department of Mathematical Sciences, University of Aarhus.

Edidin, M. (1997) Lipid microdomains in cell surface membranes. *Curr. Opin. Struct. Biol.*, **7**, 528–532.

Egorov, Yu.V. (1967) Hypoelliptic pseudodifferential operators, *Tr. Mosk. Mat. Obs.*, **16**, 99–108.

Egorov, Yu.V., Komech A.I., Shubin M.A. (2013) *Partial Differential Equations II. Elements of the Modern Theory. Equations with Constant Coefficients.* Springer.

Eidelman, S.D., Kochubei, A.N. (2004) Cauchy problem for fractional diffusion equations. *J. Differ. Equ.* **199**, 211–255.

Einstein, A. (1905) Über die von der molekularkinetischen Theorie der Wärme geforderte Bewegung von in ruhenden Flüssigkeiten suspendierten Teilchen. *Annalen der Physik,* **17** (8), 549–560. [English Translation: On the Motion of Small Particles Suspended in a Stationary Liquid, as Required by the Molecular Kinetic Theory of Heat. See online: http://users.physik.fu-berlin.de/˜ kleinert/files/eins_brownian.pdf]

Embrechts, P., Maejima, M. (2002) *Selfsimilar Processes.* Princeton University Press.

Engel, K.-J., Nagel, R. (1999) *One-parameter Semigroups for Linear Evolution Equations.* Springer.

Fannjiang, A., Komorowski, T. (2000) Fractional Brownian motions and enhanced diffusion in a unidirectional wave-like turbulence. *J. Stat. Phys.* **100** (5–6), 1071–1095.

Frank T.D. (2006) *Nonlinear Fokker-Planck equations.* Springer.

Friedman, A. (1964) *Partial Differential Equations of Parabolic Type.* Prentice-Hall.

Fokker A.D. (1914) Die mittlere Energie rotierender elektrischer Dipole im Strahlungsfeld. *Annalen der Physik,* **348** (5), 810–820.

Fujita, Y. (1990) Integrodifferential equation which interpolates the heat and the wave equations. *Osaka J. Math.* **27** 309-321, 797–804.

Gajda, J., Magdziarz, M. (2010) Fractional Fokker–Planck equation with tempered α-stable waiting times: Langevin picture and computer simulation. *Phys. Rev. E.* **82**, 011117.

Gard, T.C. (1988) *Introduction to stochastic differential equations.* Marcel Dekker, Inc.

Gardiner C.W. (1985) *Handbook of Stochastic Methods.* Springer.

Gazanfer, Ü. (2006) Fokker-Planck-Kolmogorov equation for fBM: derivation and analytical solution. *Math. Physics,* 53–60. Proceedings of the 12th regional conference, Islamabad.

Gell-Mann, M., Tsallis, C. (2004) *Nonextensive Entropy - Interdisciplinary Applications* Oxford University Press, New York.

Ghosh, R.N., Webb, W.W. (1994) Automated detection and tracking of individual and clustered cell surface low density lipoprotein receptor molecules. *Biophys. J.,* **66**, 1301–1318.

Gillis J.E., Weiss, G.H. (1970) Expected number of distinct sites visited by a random walk with an infinite variance. *J. Math. Phys.* **11**, 1307–1312.

Gorenflo, R., Mainardi, F. (1997) Fractional calculus: integral and differential equations of fractional order. In A. Carpinteri and F. Mainardi (editors): *Fractals and Fractional Calculus in Continuum Mechanics.* Springer. 223–276.

Gorenflo, R., Mainardi, F. (1998) Random walk models for space-fractional diffusion processes. *Fract. Calc. Appl. Anal.* **1** (2) 167–191.

Gorenflo, R., Mainardi, F. (1999) Approximation of Lévy-Feller diffusion by random walk. *ZAA* **18** (2) 231–246.

Gorenflo, R., Mainardi, F. (2001) Random walk models approximating symmetric space-fractional diffusion processes. In J. Elschner, I. Gohberg and B. Silbermann (editors): *Problems and Methods in Mathematical Physics.* 120–145. Operator Theory: Advances and Applications, **121**. Birkhäuser, Basel.

Gorenflo, R., Mainardi, F. (2005) Simply and multiply scaled diffusion limits for continuous time random walks. *Journal of Physics: Conference Series* **7** 1–16.

Gorenflo, R., Mainardi, F. (2008) Continuous time random walk, Mittag-Leffler waiting time and fractional diffusion: mathematical aspects. In R. Klages, G. Radons and I. M. Sokolov (editors): *Anomalous Transport: Foundations and Applications.* 93–127. Wiley-VCH, Weinheim, Germany. arXiv:0705.0797v2 [cond-mat.stat-mech].

Gorenflo, R., Luchko, Yu.F., Umarov, S. (2000) The Cauchy and multi-point partial pseudo-differential equations of fractional order. *Fract. Calc. Appl. Anal.,* **3** (3), 249–275.

Gorenflo, R., Mainardi, F., Moretti, D., Pagnini, G., Paradisi, P. (2002) Discrete random walk models for space-time fractional diffusion. *Chemical Physics,* **284**, 521–541.

Gorenflo, R., Mainardi, F., Scalas, E., Raberto, M. (2001) Fractional calculus and continuous-time finance. III. *Mathematical Finance.* 171–180. Trends Math., Birkhäuser, Basel.

Gorenflo, R., Mainardi, F., Vivoli, A. (2007) Continuous time random walk and parametric subordination in fractional diffusion. *Chaos, Solitons Fractals.* **34** (1) 87–103. arXiv:cond-mat/0701126v3 [cond-mat.stat-mech].

Gorenflo, R., Vivoli, A. (2003) Fully discrete random walks for space-time fractional diffusion equations. *Signal Processing,* **83**, 2411–2420.

Gribbin, D., Harris, R., Lau, H. (1992) Futures prices are not stable-Paretian distributed. *J. Future Markets.* **12**(4), 475–487.

Hahn, M.G., Jiang, X., Umarov, S. (2010) On q-Gaussians and exchangeability. *J. Phys. A,* **43** (16), 165208.

Hahn, M.G., Kobayashi, K., Ryvkina, J., Umarov, S.R. (2011a) On time-changed Gaussian processes and their associated Fokker-Planck-Kolmogorov equations. *Electron. Commun. Probab.* **16**, 150–164.

Hahn, M., Kobayashi, K., Umarov, S. (2012) SDEs driven by a time-changed Lévy process and their associated time-fractional order pseudo-differential equations. *J. Theoret. Probab.* **25** (1) 262–279.

Hahn, M., Kobayashi, K., Umarov, S. (2011b) Fokker–Planck–Kolmogorov equations associated with time-changed fractional Brownian motion. *Proc. Amer. Math. Soc.* **139**, 691–705.

Hahn, M.G., Umarov, S.R. (2011) Fractional Fokker-Plank-Kolmogorov type equations and their associated stochastic differential equations. *Fract. Calc. Appl. Anal.,* **14** (1), 56–79.

Hall, J., Brorsen, B.W., Irwin, S. (1989) The distribution of futures prices: A test of the stable Paretian and mixture of normals hypotheses. *J. Financ. Quant. Anal.* **24**(1), 105–116.

Hawkes, J. (1974) Local times and zero sets for processes with infinitely divisible distributions. *J. Lond. Math. Soc.* **8**, 517–525.

Hirsa, A., Neftci, S.N. (2000) *An Introduction to the Mathematics of Financial Derivatives.* Elsevier.

Hoh, W. (2000) Pseudo differential operators with negative definite symbols of variable order. *Rev. Mat. Iberoam,* **16** (2) 219–241.

Hörmander, L. (1961) Hypoelliptic differential operators. *Annales de l'Institut Fourier.* **11**, 477–492.

Hörmander, L. (1965) Pseudo-differential operators. *Comm. Pure Appl. Math.* **18**, 501–517.

Hörmander, L. (1967) Hypoelliptic second order differential equations. *Acta Math.* **119** (3-4), 147–171.

Hörmander, L. (1983) *The Analysis of Linear Partial Differential Operators. II. Differential Operators with Constant Coefficients.* Springer-Verlag, Berlin.

Hörmander, L. (2007) *The Analysis of Linear Partial Differential Operators, III,* 2nd edition. Springer.

Hu, Y., Øksendal, B. (1999) Fractional white noise calculus and applications to finance. Preprint, Department of Mathematics, University of Oslo, 10.

Ikeda, N., Watanabe, S. (1989) *Stochastic differential equations and diffusion processes.* North-Holland/Kodansha.

Jacob, N. (2001, 2002, 2005) *Pseudo-differential Operators and Markov Processes.* Vol. **I**: *Fourier Analysis and Semigroups,* Vol. **II**: *Generators and Their Potential Theory,* Vol. **III**: *Markov Processes and Applications.* Imperial College Press, London.

Jacob, N., Leopold, H.G. (1993) Pseudo differential operators with variable order of differentiation generating Feller semigroups. *Integr. Equat. Oper. Th.*, **17**, 544–553.

Jacob, N., Krägeloh, A. (2002) The Caputo derivative, Feller semigroups, and the fractional power of the first order derivative on $C_\infty(R_0^+)$. *Fract. Calc. and Appl. Anal.*, **5**, (4) 395–410.

Jacod, J. (1979) *Calcul Stochastique et Problèmes de Martingales.* Lecture Notes in Mathematics, **714**. Springer, Berlin.

Jacod, J., Shiryaev, A.N. (1987) *Limit Theorems for Stochastic Processes.* Springer-Verlag, Berlin, Heidelberg, New York.

Jakubowski, A., Mémin, J., Pagès, G. (1989) Convergence en loi des suites d'intégrales dur l'espace D^1 de Skorokhod. *Probab. Theory Related Fields*, **81** (1), 111–137.

Janczura, J., Wiłomańska, A. (2009) Subdynamics of financial data from fractional Fokker–Planck equation. *Acta Phys. Pol. B.* **40**, 1341–1351.

Janczura, J., Wiłomańska, A. (2012) Anomalous diffusion models: different types of subordinator distribution. *Acta Phys. Pol.* **43** (5).

Janson, S. (1997) *Gaussian Hilbert Spaces.* Cambridge University Press.

Jum, E., Kobayashi, K. (2016) A strong and weak approximation scheme for stochastic differential equations driven by a time-changed Brownian motion. *Probab. Math. Statist.* **36** (2), 201–220.

Kallsen, J., Shiryaev, A.N. (2002) Time change representation of stochastic integrals. *Theory Probab. Appl.* **46**, 522-528.

Karatzas, I., Shreve, S.E. (1991) *Brownian Motion and Stochastic Calculus.* Springer.

Kilbas, A.A., Srivastava, H.M., Trujillo, J.J. (2006) *Theory and applications of fractional differential equations.* Elsevier.

Kloeden, P.E., Platen, E. (1999) *Numerical Solution of Stochastic Differential Equations.* Springer.

Kobayashi, K. (2011) Stochastic calculus for a time-changed semimartingale and the associated stochastic differential equations. *J. Theoret. Probab.*, **24** (3), 789–820.

Kochubei, A.N. (2008) Distributed order calculus and equations of ultraslow diffusion. *J. Math. Anal. Appl.* **340**, 252–281.

Kohn, J.J., Nirenberg, L. (1965) An algebra of pseudo-differential operators. *Comm. Pure Appl. Math.* **18** 269–305.

Kolokoltsov, V.N. (2009) Generalized continuous-time random walk (CTRW), subordinating by hitting times and fractional dynamics. *Theor. Prob. Appl.* **53** (4).

Kolmogorov, A.N. (1931) Über die analytischen Methoden in der Wahrscheinlichkeitsrechnung. *Mathematische Annalen,* **104** (1), 415–458.

Kon, S. (1984) Models of Stock Returns - A Comparison. *Journal of Finance*, **39**, 147–165.

Kovács, M., Meerschaert, M.M. (2006) Ultrafast subordinators and their hitting times. *Publ. de l'Institut Mathématique* Novelle série, **80**(94), 193–206.

Kumar, A., Vellaisamy, P. (2015) Inverse tempered stable subordinators. *Stat. Probab. Lett.*, **103**, 134–141.

Kurtz, T.G, Protter, P. (1991a) Characterizing the weak convergence of stochastic integrals. In *Stochastic Analysis: Proceedings of the Durham Symposium on Stochastic Analysis* (Barlow, M.T. Bingham, N.H., editors), London Math Society Lecture Note Series, **167**, 225–259, Cambridge University Press.

Kurtz, T.G, Protter, P. (1991b) Weak limit theorems for stochastic integrals and stochastic differential equations. *Ann. Probab.*, **19**, 1035–1070.

Kurtz, T.G, Protter, P. (1996) Weak convergence of stochastic integrals and differential equations. In *Probabilistic Models for Nonlinear Partial Differential Equations* (Talay, D. Tubaro, L., editors), Lecture Notes in Mathematics, **1627**. 1–41, Springer-Verlag.

Kyprianou, A.E. (1992) *Introductory Lectures on Fluctuations of Levy Processes with Applications*. Springer.

Langevin, P. (1908) Sur la théorie do mouvement brownien. *C. R. Acad. Sci.* (Paris), **146**, 530–533 [Enlish Transl: On the theorie of Brownian motion. Available online http://www.physik.uni-augsburg.de/theo1/hanggi/History/Langevin1908.pdf]

Liu, F., Shen, S., Anh, V.V., Turner, I. (2004) Analysis of a discrete non-Markovian random walk approximation for the time fractional diffusion equation. *ANZIAM J. (E)*, **46**, 488–504.

Liu B., Goree J. (2008) Superdiffusion and non-Gaussian statistics in a driven-dissipative 2D dusty plasma. *Phys. Rev. Lett.* **100**, 055003.

Lipster, R.S., Shiryaev, A.N. (1989) *Theory of Martingales*. Kluwer Acad. Publ. Dordrecht.

Lorenzo, C.F., Hartley, T.T. (2002) Variable order and distributed order fractional operators. *Nonlinear Dynamics*, **29**, 57–98.

Lutz, E. (2003) Anomalous diffusion and Tsallis statistics in an optical lattice. *Phys. Rev.*, A **67**.

Lv, L., Qiu, W., Ren, F. (2012) Fractional Fokker-Planck equation with space and time dependent drift and diffusion. *J. Stat. Phys.* **149**, 619–628.

Magdziarz, M. (2009a) Langevin picture of subdiffusion with infinitely divisible waiting times. *J. Stat. Phys.* **135**, 763–772.

Magdziarz, M. (2009b) Stochastic representation of subdiffusion processes with time-dependent drift. *Stoch. Proc. Appl.* **119**, 3238–3252.

Magdziarz, M. (2009c) Black-Scholes formula in subdiffusive regime. *J. Stat. Phys.* **136**, 553–564.

Magdziarz, M., Gajda, J., Zorawik, T. (2014) Comment on fractional Fokker-Planck equation with space and time dependent drift and diffusion. *J. Stat. Phys.* **154**, 1241–1250.

Magdziarz, M., Weron, A. (2007) Competition between subdiffusion and Lévy flights: a Monte Carlo approach. *Phys. Rev. E* **75**, 056702.

Magdziarz, M., Weron, A., Klafter, J. (2008) Equivalence of the fractional Fokker-Planck and subordinated Langevin equations: the case of a time-dependent force. *Phys. Rev. Lett.* **101**, 210601.

Mainardi, F. (2010) *Fractional Calculus and Waves in Linear Viscoelasticity: An Introduction to Mathematical Models*. Imperial College Press.

Mainardi, F., Luchko, Y., Pagnini, G. (2001) The fundamental solution of the space-time fractional diffusion equation. *Fract. Calc. Appl. Anal.* **4** (2), 153–192.

Mainardi, F., Raberto, M., Gorenflo, R., Scalas, E. (2000) Fractional calculus and continuous-time finance II: the waiting-time distribution. *Physica A,* **287** (3) 468–481.

Malliavin, P. (1997) *Stochastic Analysis.* Springer-Verlag.

Mandelbrot, B. (1997) *Fractals and Scaling in Finance.* Springer.

McCauley, J.L., Gunaratne, G.H., Bassler, K.E. (2006) Hurst exponents, Markov processes and fractional Brownian motion. *MPRA.*

McCulloch, J.H. (1996) Financial applications of stable distributions. *In G. S. Maddala and C. R. Rao (Eds.), Handbook of Statistics.* **14**. New York: North-Holland.

Meerschaert, M.M., Benson, D. (2002) Governing equations and solutions of anomalous random walk limits. *Phys. Rev. E,* **66**, 060102(R), 1–4.

Meerschaert, M.M., Benson, D., Bäumer, B. (2001) Operator Lévy motion and multiscaling anomalous diffusion. *Phys. Rev. E* **63**, 021112–021117.

Meerschaert, M.M., Benson, D.A., Scheffler, H-P., Baeumer, B. (2002a) Stochastic solution of space-time fractional diffusion equations. *Phys. Rev. E.* **65**, 1103–1106.

Meerschaert, M.M., Benson, D.A., Scheffler, H-P., Becker-Kern, P. (2002b) Governing equations and solutions of anomalous random walk limits. *Phys. Rev. E.* **66** (6), 102R–105R.

Meerschaert M.M., Nane, E., Xiao, Y. (2008) Large deviations for local time fractional Brownian motion and applications, *J. Math. Anal. Appl.* **346**, 432–445.

Meerschaert, M.M., Nane, E., Xiao Y. (2009a) Correlated continuous time random walks. *Stat. Probab. Lett.* **79**, 1194-1202.

Meerschaert M.M., Nane E., Vellaisamy P. (2009b) Fractional Cauchy problems on bounded domains. *Ann. Proba.* **37**, 3, 979–1007.

Meerschaert, M.M., Scheffler, H-P. (2001) *Limit Distributions for Sums of Independent Random Vectors. Heavy Tails in Theory and Practice.* John Wiley and Sons, Inc.

Meerschaert, M.M., Scheffler, H.-P. (2004) Limit theorems for continuous-time random walks with infinite mean waiting times. *J. Appl. Probab.* **41**, 623–638.

Meerschaert, M.M., Scheffler H.-P. (2005) Limit theorems for continuous time random walks with slowly varying waiting times. *Stat. Probab. Lett.* **71** (1), 15–22.

Meerschaert, M.M., Scheffler, H.-P. (2006) Stochastic model for ultraslow diffusion. *Stoch. Proc. Appl.* **116**, 1215–1235.

Meerschaert, M.M., Scheffler, H.-P. (2008) Triangular array limits for continuous time random walks. *Stoch. Proc. Appl.* **118**, 1606–1633.

Mendes, V.R. (2009) A fractional calculus interpretation of the fractional volatility model. *Nonlinear Dynamics.* **55**: 395.

Merton, R. (1973) Theory of rational option pricing. *Bell J. Econom. Maneg. Sci. The RAND Corporation.* **4** (1), 141–183.

Metzler, R., Barkai, E., Klafter, J. (1999) Anomalous diffusion and relaxation close to thermal equilibrium: a fractional Fokker-Planck equation approach. *Phys. Rev. Lett.* **82**, 3563–3567.

Metzler, R., Klafter, J. (2000) The random walk's guide to anomalous diffusion: a fractional dynamics approach. *Phys. Rep.* **339** (1), 1–77.

McCauley, J.L. (2013) *Stochastic Calculus and Differential Equations for Physics and Finance.* Cambridge Books Online.

Miao, Y., Ren, W., Ren, Z. (2008) On the fractional mixed fractional Brownian motion.

Appl. Math. Sci. **35**, 1729–1938.

Mittnik, S., Rachev, S. (1993) Modeling asset returns with alternative stable distributions. *Economics Reviews* **12**, 261–330.

Monroe, I. (1978) Processes that can be embedded in Brownian motion. *Ann. Probab.* **6** (1) 42–56.

Montroll E.W., Weiss G.H. (1965) Random walks on lattices. II. *J. Math. Phys.* **6**, 167–181.

Montroll, E.W., Shlesinger M.F. (1982) On $1/f$ noise and other distributions with long tails. *Proc. Nat. Acad. Sci. USA* **79**, 3380–3383.

Montroll, E.W., Bendler, J.T. (1984) On Lévy (or stable) distributions and the Williams-Watt model of dielectric relaxation. *J. Stat. Phys.* **34**, 129–162.

Nane, E., Ni, Y. (2017) Stability of the solution of stochastic differential equation driven by time-changed Lévy noise. *Proc. Amer. Math. Soc.* **145**, 3085–3104

Nigmatullin, R.R. (1986) The realization of generalized transfer equation in a medium with fractal geometry. *Phys. Stat. Sol. B* **133**, 425–430.

Nualart, D. (2006) *The Malliavin Calculus and Related Topics, 2nd edition.* Springer.

Oldham, K., Spanier, J. (1974) *The Fractional Calculus: Theory and Application of Differentiation and Integration to Arbitrary Order.* Acad. Press, Dover Publications, New York - London.

Osu, B.O., Ifeoma, C.A. (2016) Fractional Black Scholes Option Pricing with Stochastic Arbitrage Return. *International Journal of Partial Differential Equations and Applications,* **4** (2), 20–24.

Pierce, R.D. (1996) RCS characterization using the alpha-stable distribution. *In Proceedings of the National Radar Conference,* IEEE Press, 154–159.

Planck, M. (1917) Über einen Satz der statistischen Dynamik und seine Erweiterung in der Quantentheorie. *Sitzungsber. Preuss. Akad. Wiss.,* **24**.

Podlubny, I. (1998) *Fractional Differential Equations.* Mathematics in Science and Engineering, **198**. Academic Press.

Prato, D., Tsallis, C. (1999) Nonextensive foundation of Lévy distributions. *Phys. Rev. E* **60**, 2398–2401.

Prokhorov, Yu.V. (1956) Convergence of random processes and limit theorems in probability. *Theor. Prob. Appl.,* **1**, 157–214.

Protter, P. (2004) *Stochastic Integration and Differential Equations,* 2nd ed., Springer.

Pruitt, W.E. (1969) The Hausdorff dimension of the range of a process with stationary independent increments. *J. Math. Mech.* **19**, 371–378.

Rasmussen, C.E., Williams, C.K.I. (2006) *Gaussian Processes for Machine Learning.* MIT Press.

Revuz, D., Yor, M. (1999) *Continuous Martingales and Brownian Motion,* 3rd edition. Grundlehren der Mathematischen Wissenschaften, **293**. Springer, Berlin.

Risken, H., Frank, T. (1996) *The Fokker-Planck Equation: Methods of Solutions and Applications.* Springer.

Ritchie, K., Shan, X.-Y., Kondo, J., Iwasawa, K., Fujiwara, T., Kusumi, A. (2005) Detection of non-Brownian diffusion in the cell membrane in single molecule tracking. *Biophys. J.,* **88**, 2266–2277.

Rodriguez, A., Schwammle, V., Tsallis, C. (2008) Strictly and asymptotically scale invariant probabilistic models of N correlated binary random variables having q-Gaussians as $N \to \infty$ limiting distributions. *J. Stat. Mech.,* P09006.

Rogers, L.C.G. (1997) Arbitrage with fractional Brownian motion. *Mathematical Finance.* **7**, 95–105.

Rosiński, J. (2007) Tempering stable processes. *Stoch. Proc. Appl.*, **117** (6), 677–707.

Rosiński, J. (2001) Series representations of Lévy processes from the perspective of point processes. In *Lévy Processes: Theory and Applications.* Barndorff-Nielsen, O.E., Mikosch, T., Resnick, S.I. Eds. 401–415. Birkhäuser, Boston.

Rubin, B. (1996) *Fractional Integrals and Potentials.* Pitman Monographs and Surveys in Pure and Applied Math. **82**, Longman.

Samko, S.G., Kilbas, A.A., Marichev, O.I. (1993) *Fractional Integrals and Derivatives: Theory and Applications.* Gordon and Breach Science Publishers.

Samorodnitsky, G., Taqqu, M.S. (1994) *Stable non-Gaussian Random Processes.* Chapman & Hall, New York.

Sato, K-i. (1999) *Lévy Processes and Infinitely Divisible Distributions.* Cambridge University Press.

Saxton, M.J. (2001) Anomalous Subdiffusion in Fluorescence Photobleaching Recovery: A Monte Carlo Study. *Biophys. J.*, **81**(4), 2226–2240.

Saxton, M.J., Jacobson, K. (1997) Single particle tracking: Applications to membrane dynamics. *Ann. Rev. Biophys. Biomol. Struct.*, **26**, 373–399.

Scalas, E., Gorenflo, R., Mainardi, F. (2000) Uncoupled continuous-time random walks: Solution and limiting behavior of the master equation. *Phys. Rev. E*, **69** (1), 011107.

Scalas, E., Viles, N. (2013) A functional limit theorem for stochastic integrals driven by a time-changed symmetric α-stable Lévy process. *Stoch. Proc. Appl.* **124** (1), 385–410.

Schertzer, D., Larchevêque, M., Duan, J., Yanovsky, V.V., Lovejoy, S. (2001) Fractional Fokker-Planck equation for nonlinear stochastic differential equations driven by non-Gaussian Lévy stable noises. *J. Math. Phys.* **42** (1) 200–212.

Seneta, E. (1976) *Regularly Varying Functions.* Lecture Notes in Mathematics, **508** Springer Berlin Heidelberg.

Shiryaev, A. (1999) *Essentials of Stochastic Finance: Facts, Models, Theory.* World Scientific Publishing.

Schmitt, F.G., Seuront, L. (2001) Multifractal random walk in copepod behavior. *Physica A*, **301**, 375–396.

Schneider, W.R. (1990) Fractional diffusion. *Lect. Notes Phys.* **355**, Heidelberg, Springer, 276–286.

Schneider, W.R., Wyss, W. (1989) Fractional diffusion and wave equations. *J. Math. Phys.* **30**, 134–144.

Schoutens, W. (2003) *Lévy processes in Finance.* John Wiley & Sons.

Schutz, G.J., Schindler, H., Schmidt, T. (1997) Single-molecule microscopy on model membranes reveals anomalous diffusion. *Biophys. J.*, **73** (2), 1073–1080.

Schulze B.-W. (1991) *Pseudo-Differential Operators on Manifolds with Singularities. Volume 24 of Studies in Math. and Appl.* Elsevier.

Sheng-Wu, H., Jia-Gang, W., Jia-An, Y. (1992) *Semimartingale Theory and Stochastic Calculus.* Science Press.

Shubin, M.A. (2001) *Pseudodifferential Operators and Spectral Theory.* Springer.

Shizgal, B. (2015) *Spectral Methods in Chemistry and Physics. Applications to Kinetic Theory and Quantum Mechanics.* Springer.

Silbergleit, V.M., Gigola, S.V., D'Attellis C.E. (2007) Statistical studies of sunspots. *Acta Geodaetica et Geophysica Hungarica*, **49** (3), 278–283.

Silvestrov, D.S. (2004) *Limit Theorems for Randomly Stopped Stochastic Processes.* Springer.

Situ, R. (2005) *Theory of Stochastic Differential Equations with Jumps and Applications: Mathematical and Analytical Techniques with Applications to Engineering.* Springer.

Skorokhod, A.V. (1956) Limit theorems for stochastic processes. *Theor. Prob. Appl.* **1**, 261-290.

Slattery, J.P. (1995) *Lateral mobility of FceRI on rat basophilic leukemia cells as measured by single particle tracking using a novel bright fluorescent probe.* Ph.D. thesis, Cornell University, Ithaca, NY.

Smoluchowski, M. (1906) Zur kinetischen Theorie der Brownschen Molekularbewegung und der Suspensionen. *Annalen der Physik*, **21** (14), 756–780.

Sokolov, I.M., Blumen, A., Klafter, J. (2001) Dynamics of annealed systems under external fields: CTRW and the Fractional Fokker-Planck equations. *Europhys. Lett*, **56**(2), 175–180.

Sokolov, I.M., Chechkin, A.V., Klafter, J. (2004) Distributed-order fractional kinetics. *Acta Physica Polonica B*, **35**, 1323–1341.

Sokolov, I.M., Klafter, J. (2006) Field-induced dispersion in subdiffusion. *Phys. Rev. Lett.* **97**, 140602.

Song, L., Wang, W. (2013) Solution of the fractional Black-Scholes option pricing model by finite difference method. *Abstract and Appl. Anal.* Article ID 194286. Available online http://dx.doi.org/10.1155/2013/194286

Sottinen, T., Valkeila, E. (2003) On arbitrage and replication in the fractional BlackScholes pricing model. *Statistics and Decisions.* **21**, 137–151.

Steele, J.M. (2001) *Stochastic Calculus and Financial Applications.* Applications of Mathematics (New York), **45**. Springer.

Stroock, D.W. (2003) *Markov Processes from K. Itô's perspective.* Princeton University Press.

Suzuki, K., Ritchie, K., Kajikawa, E., Fujiwara, T., Kusumi, A. (2005) Rapid hop diffusion of a G-protein-coupled receptor in the plasma membrane as revealed by single-molecule techniques. *Biophys. J.* **88** (5), 3659–3680.

Thäle, C. (2009) Further remarks on mixed fractional Brownian motion. *Appl. Math. Sci.* **3** (38), 1885–1901.

Taira, K. (1992) On the existence of Feller semigroups with boundary conditions. *Memoirs of AMS*, **99** (475).

Taira, K. (2004) *Semigroups, Boundary Value Problems and Markov Processes.* Springer.

Taylor, M. (1981) *Pseudodifferential Operators.* Princeton University Press.

Tsallis, C., Bukman, D.J. (1996) Anomalous diffusion in the presence of external forces: exact time-dependent solutions and their thermostatistical basis. *Phys. Rev. E* **54**, R2197.

Triebel, H. (1977) *Interpolation Theory, Function Spaces, Differential Operators.* Birkhäuser, Basel.

Tsallis C. (2009) *Introduction to Nonextensive Statistical Mechanics: Approaching a Complex World.* Springer.

Uchaykin, V.V., Zolotarev V.M. (1999) *Chance and Stability. Stable Distributions and their Applications.* VSP, Utrecht.

Umarov, S. (2015a) Continuous time random walk models associated with distributed order diffusion equations. *Frac. Calc. Appl. Anal.* **18** (3), 821–837.

Umarov, S. (2015b) *Introduction to Fractional and Pseudo-differential Equations with Singular Symbols.* New York, Springer.

Umarov, S., Gorenflo, R. (2005a) Cauchy and nonlocal multi-point problems for distributed order pseudo-differential equations. I. *Z. Anal. Anwend.* **24** (3) 449–466.

Umarov, S., Gorenflo, R. (2005b) On multi-dimensional random walk models approximating symmetric space-fractional diffusion processes. *Fract. Calc. Appl. Anal.*, **8**, 73–88.

Umarov, S., Steinberg, S. (2006) Random walk models associated with distributed fractional order differential equations. *IMS Lecture Notes - Monograph Series. High Dimensional Probability.* **51**, 117–127.

Umarov, S., Steinberg, S. (2009) Variable order differential equations with piecewise constant order-function and diffusion with changing modes. *Z. Anal. Anwend.* **28**, 431–450.

Umarov, S., Tsallis, C., Steinberg, S. (2008) On a q-central limit theorem consistent with nonextensive statistical mechanics. *Milan Journal of Mathematics*, **76**, 307.

Umarov, S., Tsallis, C., Gell-Mann, M., Steinberg. S. (2010) Generalization of symmetric alpha-stable distributions to $q > 1$. *J. Math. Phys.*

Umarov, S., Saydamatov, E.M. (2006) A fractional analog of the Duhamel principle. *Fract. Calc. Appl. Anal.*, **9** (1) 57–70.

Umarov, S., Saydamatov, E.M. (2007) A generalization of the Duhamel principle for fractional order differential equations. *Doclady Ac. Sci. Russia*, **412** (4) 463–465.

Wang, X.-T. (2010) Scaling and long-range dependence in option pricing I: Pricing european option with transaction costs under the fractional BlackScholes model. *Physica A: Statistical Mechanics and its Applications*, **389**, 438–444.

Wang, X.-T., Zhu, E.-H., Tang, M.-M., and Yan, H.-G. (2010) Scaling and long-range dependence in option pricing II: Pricing european option with transaction costs under the mixed brownian fractional brownian model. *Physica A: Statistical Mechanics and its Applications*, **389**, 445–451.

Wentcel, A.D. (1959). On boundary conditions for multidimensional diffusion processes. *Theory. Probab. Appl.* **4**, 164–177.

Whitt, W. (2002) *Stochastic-process Limits: an Introduction to Stochastic-Process Limits and their Application to Queues.* Springer.

Widder, D.V. (1941) *The Laplace transform.* Princeton University Press.

Wolf, J. (2010) Random Fractals Determined by Lévy processes. *J. Theor. Probab.*, **23** (4), 1182–1203.

Wu, Q. (2016) Stability of stochastic differential equations with respect to time-changed Brownian motions. `arXiv:1602.08160` [math.PR]

Wyss, W. (1986) The fractionnal diffusion equation. *J. Math. Phys.* **27**, 2782–2785.

Wyss, W. (2000) The fractional Black-Scholes equation. *Fract. Calc. Appl. Anal.* **3** (1), 51–61.

Xiao, Y. (2004) Random fractals and Markov processes. In *Fractal geometry and applications: a jubilee of Benoit Mandelbrot, Part 2.* Proc. Sympos. Pure Math, **72**,

Amer. Math.Soc., Providence, RI, 261-338. Updated version obtained from author's website.

Zaslavsky, G (2002) Chaos, fractional kinetics, and anomalous transport. *Phys. Rep.*, **371**, 461–580.

Index

Printed in the United States
By Bookmasters